JAAVSO

The Journal of
The American Association
of Variable Star Observers

Volume 46
Number 2
2018

AAVSO
49 Bay State Road
Cambridge, MA 02138
USA

ISSN 0271-9053 (print)
ISSN 2380-3606 (online)

Publication Schedule

The Journal of the American Association of Variable Star Observers is published twice a year, June 15 (Number 1 of the volume) and December 15 (Number 2 of the volume). The submission window for inclusion in the next issue of JAAVSO closes six weeks before the publication date. A manuscript will be added to the table of contents for an issue when it has been fully accepted for publication upon successful completion of the referee process; these articles will be available online prior to the publication date. An author may not specify in which issue of JAAVSO a manuscript is to be published; accepted manuscripts will be published in the next available issue, except under extraordinary circumstances.

Page Charges

Page charges are waived for Members of the AAVSO. Publication of unsolicited manuscripts in JAAVSO requires a page charge of US $100/page for the final printed manuscript. Page charge waivers may be provided under certain circumstances.

Publication in *JAAVSO*

With the exception of abstracts of papers presented at AAVSO meetings, papers submitted to JAAVSO are peer-reviewed by individuals knowledgable about the topic being discussed. We cannot guarantee that all submissions to JAAVSO will be published, but we encourage authors of all experience levels and in all fields related to variable star astronomy and the AAVSO to submit manuscripts. We especially encourage students and other mentees of researchers affiliated with the AAVSO to submit results of their completed research.

Subscriptions

Institutions and Libraries may subscribe to JAAVSO as part of the Complete Publications Package or as an individual subscription. Individuals may purchase printed copies of recent JAAVSO issues via Createspace. Paper copies of JAAVSO issues prior to volume 36 are available in limited quantities directly from AAVSO Headquarters; please contact the AAVSO for available issues.

Instructions for Submissions

The *Journal of the AAVSO* welcomes papers from all persons concerned with the study of variable stars and topics specifically related to variability. All manuscripts should be written in a style designed to provide clear expositions of the topic. Contributors are encouraged to submit digitized text in MS WORD, LATEX+POSTSCRIPT, or plain-text format. Manuscripts may be mailed electronically to journal@aavso.org or submitted by postal mail to JAAVSO, 49 Bay State Road, Cambridge, MA 02138, USA.

Manuscripts must be submitted according to the following guidelines, or they will be returned to the author for correction:

Manuscripts must be:
    1) original, unpublished material;
    2) written in English;
    3) accompanied by an abstract of no more than 100 words.
    4) not more than 2,500–3,000 words in length (10–12 pages double-spaced).

Figures for publication must:
    1) be camera-ready or in a high-contrast, high-resolution, standard digitized image format;
    2) have all coordinates labeled with division marks on all four sides;
    3) be accompanied by a caption that clearly explains all symbols and significance, so that the reader can understand the figure without reference to the text.

Maximum published figure space is 4.5" by 7". When submitting original figures, be sure to allow for reduction in size by making all symbols, letters, and division marks sufficiently large.

Photographs and halftone images will be considered for publication if they directly illustrate the text.

Tables should be:
    1) provided separate from the main body of the text;
    2) numbered sequentially and referred to by Arabic number in the text, e.g., Table 1.

References:
    1) References should relate directly to the text.
    2) References should be keyed into the text with the author's last name and the year of publication, e.g., (Smith 1974; Jones 1974) or Smith (1974) and Jones (1974).
    3) In the case of three or more joint authors, the text reference should be written as follows: (Smith et al. 1976).
    4) All references must be listed at the end of the text in alphabetical order by the author's last name and the year of publication, according to the following format: Brown, J., and Green, E. B. 1974, *Astrophys. J.*, **200**, 765.
        Thomas, K. 1982, *Phys. Rep.*, **33**, 96.
    5) Abbreviations used in references should be based on recent issues of JAAVSO or the listing provided at the beginning of *Astronomy and Astrophysics Abstracts* (Springer-Verlag).

Miscellaneous:
    1) Equations should be written on a separate line and given a sequential Arabic number in parentheses near the right-hand margin. Equations should be referred to in the text as, e.g., equation (1).
    2) Magnitude will be assumed to be visual unless otherwise specified.
    3) Manuscripts may be submitted to referees for review without obligation of publication.

Online Access

Articles published in JAAVSO, and information for authors and referees may be found online at: https://www.aavso.org/apps/jaavso/

The Journal of the American Association of Variable Star Observers
Volume 46, Number 2, 2018

Table of Contents continued on following pages

Table of Contents continued on next page

Editorial

Variable Stars: The View from Up Here

John R. Percy
Editor-in-Chief, *Journal of the AAVSO*

Department of Astronomy and Astrophysics, and Dunlap Institute for Astronomy and Astrophysics, University of Toronto, 50 St. George Street, Toronto, ON M5S 3H4, Canada; john.percy@utoronto.ca

Received November 16, 2018

Last year (2017), Canada celebrated its 150th birthday. This year (2018), the Royal Astronomical Society of Canada (RASC) does likewise. In recent *AAVSO Newsletters*, I've written short articles about both Canadian astronomy (April 2017) and about the RASC (January 2018). In this article, I will complete the trilogy with a brief overview of variable star astronomy in Canada. I'm only skimming the surface; there are many more people and achievements which deserve mentioning.

This article is also appropriate because, on June 13–16, 2019, the AAVSO Spring Meeting will be held in Toronto, jointly with the RASC annual General Assembly. AAVSO and RASC previously met together in 1940, 1957, 1961, 1965, 1974, 1983, 1999, and 2007, so this joint meeting is certainly due. It was at the 1974 joint meeting in Winnipeg that I first met former AAVSO Director Janet Mattei, and we soon became collaborators and good friends.

RASC is exemplary in its balance between local and national activities, notably its national publications. The *Journal of the RASC* (*JRASC*) was the "voice" of the AAVSO for decades: in addition to publishing general notes on variable star astronomy for over a century, *JRASC* published bi-monthly "Variable Star Notes" from AAVSO Directors Margaret Mayall and Janet Mattei from 1952 to 1981, and reported on AAVSO meetings from 1937 to 1946 and from 1952 to 1966. That's one of the ways that I first learned about the AAVSO!

RASC's annual *Observers Handbook* has always included several pages on variable stars and variable star observing, contributed by AAVSO. It includes basic information on variable stars, beginners' charts, predictions for periodic variables, and short essays on variables of special interest (for 2019, it's Nova Circini 2018). The RASC website also has useful information for potential variable star observers: www.rasc.ca/vs-overview We do not have a "variable star section"; we encourage our variable star observers to contribute through the AAVSO.

Amateur Variable Star Astronomy

We must surely begin with the enigmatic Joseph Miller Barr (1856–1911), from St. Catherines, Ontario (Percy 2015). He published papers on variable and binary stars in journals including the *Astrophysical Journal*, and has an astronomical "effect" named after him (the "Barr effect" is an apparent non-random distribution in the orientation of spectroscopic binary star orbits, probably caused by the distortion of spectral lines by gas flows in the system). But why did he never ever appear at any astronomical meeting? Was he disabled? Female? Or just reclusive? Bert Petrie was an early AAVSO observer (observer code PER) when he was an undergraduate at the University of British Columbia, using a telescope loaned by AAVSO. He published "Variable Star Observing for Amateurs" at the age of 20 (Petrie 1926), and made 137 visual observations before going on to become one of Canada's most eminent professional astronomers.

David Rosebrugh was born in Canada, but moved to the US, and became an AAVSO "star": a prolific observer, author, Secretary (1937–1945), and President (1948–1949), but remained a lifelong member of RASC.

In mid-century, Montreal became a hotbed of observational activity, including variable star observing, led especially by Isabel Williamson. This produced three AAVSO presidents: Frank de Kinder (1967–1969), Charles Good (1971–1973), and George Fortier (1975–1977). It also produced David Levy, one of the best-known amateur astronomers in the world.

Across Canada, visual variable star observers have racked up significant totals: Warren Morrison (197,712), Steven Sharpe (117,139), followed by Miroslav Komorous, Patrick Abbott, Richard Huziak, Christopher Spratt, Daniel Taylor, Patrick McDonald, Bernard Bois, and Raymond Thompson. Ray Thompson went on to become Canada's leading PEP observer with 8,231 observations; he ranks third among individual observers, all-time, world-wide. George Fortier was a pioneer in this field. Steven Sharpe is also our leading DSLR observer, with 9,533 observations. Vance Petriew and Richard Huziak have taken advantage of dark Saskatchewan skies to amass 311,973 and 144,378 CCD observations, respectively. Michael Cook, Walter MacDonald, Damien Lemay, and David Lane round out the list of those who have made over 10,000 CCD observations. Warren Morrison was also discoverer of Nova Cyg 1978, and the 1985 outburst of the recurrent nova RS Oph. Several Canadian amateurs have been involved in recent supernova discovery projects, including 10-year-olds Kathryn and Nathan Gray (with much attention from the media!). Paul Boltwood was deservedly known for his significant contributions to both hardware and software, and his application of these to photometry of active galactic nuclei. On the solar side: current AAVSO sunspot group leader Kim Hay has made over 2,500 solar observations.

Professional Variable Star Astronomy

Canada has produced its share of professional variable star astronomers, despite being climatically underprivileged. Most notable is Helen Sawyer Hogg. She was born and educated in the US, but spent most of her career (1935–1993) at the University of Toronto. She was a pioneer woman in the physical sciences, an internationally-recognized researcher on variable stars in globular clusters, and a weekly columnist on astronomy in the *Toronto Star* (Canada's largest-circulation newspaper) for over 30 years. She served as President of the AAVSO (1939–1941), and of the RASC (1957–1959), and was founding President of the Canadian Astronomical Society, our professional organization, in 1971–1972. In 1976, she was appointed Companion of the Order of Canada, the highest rank in the Order—our equivalent to knighthood.

Equally eminent, though not active in the AAVSO or RASC is Sidney van den Bergh, honoured for his research on stars and galaxies, including supernovae, Cepheids, the period-luminosity (Leavitt) relation, and the extragalactic distance scale.

My colleague Don Fernie served as President of RASC (1974–1976), as well as President of the International Astronomical Union's (IAU) Commission on Variable Stars. He was a photometrist who published widely on Cepheids and other pulsating stars, and also on the history of astronomy. In the 1960s and 1970s, he and his colleagues (including me) took advantage of the long-term availability of the David Dunlap Observatory's spectroscopic and photometric facilities to supervise a series of landmark doctoral thesis projects on the long-term behavior of variable stars: Mira stars (Tom Barnes, Richard Crowe, Nancy Remage Evans), RR Lyrae stars (Christine Coutts Clement), Classical Cepheids (Nancy Remage Evans, Robert Gauthier), Population II Cepheids (Serge Demers), RV Tauri stars (David DuPuy), other yellow supergiants (Armando Arellano), RCB stars (Vicki Watt), RS CVn stars (Dorothy Fraquelli, Bill Herbst), other eclipsing binaries (Paul Hendry). There were other variable star theses, based on purely spectroscopic observations of, for example, binary and peculiar stars, or obtained with other facilities such as the University of Toronto Southern Observatory in Chile; and, although Toronto has been Canada's most prolific "variable star factory," other universities across the country have contributed, also.

Jaymie Matthews was the public "face" of Canada's MOST (Microvariability and Oscillations of STars) variable star satellite, though Slavek Rucinski was the "brains" behind MOST and its successor, the BRITE constellation of variable star nanosatellites. Jaymie is currently President of the IAU Commission on Pulsating Stars.

And let's not forget Ian Shelton, the primary discoverer of Supernova 1987A, Peter Stetson who developed DAOPHOT—one of the most useful tools for CCD photometry of variable stars—and Arthur Covington, who pioneered the radio study of variable solar activity, starting after WWII. Peter Millman, a world expert on meteors, published several papers on variable stars, served on AAVSO Council from 1947 to 1949 and from 1958 to 1960, and was a strong supporter of "citizen science." Doug Welch is an example of a professional who began as a keen amateur. He served three terms on AAVSO Council, was an advisor on several AAVSO projects, and to the short-period pulsator section, and has contributed to many areas of variable star research, including Cepheids, RR Lyrae stars, RCB stars, supernovae and their light echoes. He is currently a Dean and Vice-Provost at McMaster University.

There are also expatriate Canadians such as Wendy Freedman, co-recipient of the Gruber Cosmology Prize for her work in using HST and Cepheids to establish the extragalactic distance scale, and the age of the universe; and David Charbonneau and Sara Seager, two of the leaders in studying the nature and properties of exoplanets, including through their transits. In Canada, interest in variable stars—including exotic kinds—continues, across the country. My colleagues are leaders in the study of things like supernovae, pulsars, X-ray bursters, and the latest mystery, fast radio bursts—as well as a few of us who continue to study less-exotic types of variables. And our amateur observers are as busy and productive as ever!

Acknowledgements

I thank Elizabeth Waagen for providing up-to-date statistics on Canadian observers' totals, and Elizabeth and Mike Saladyga for reading a draft version of this editorial.

References

Percy, J. R. 2015, *J. Roy. Astron. Soc. Canada*, **109**, 270.
Petrie, R. M. 1926, *J. Roy. Astron. Soc. Canada*, **20**, 42.

John Percy is a variable star astronomer who has served as President (1989–1991) of the AAVSO, and President (1978–1980) and Honorary President (2013–2018) of the RASC. He has received both the Merit Award and the William Tyler Olcott Award from the AAVSO.

Period Study and Analysis of 2017 BVR$_c$I$_c$ Observations of the Totally Eclipsing, Solar Type Binary, MT Camelopardalis

Ronald G. Samec
Faculty Research Associate, Pisgah Astronomical Research Institute, 1 PARI Drive, Rosman, NC 28772; ronaldsamec@gmail.com

Daniel B. Caton
Dark Sky Observatory, Department of Physics and Astronomy, Appalachian State University, 525 Rivers Street, Boone, NC 28608

Danny R. Faulkner
Johnson Observatory, 1414 Bur Oak Court, Hebron, KY 41048

Received June 8, 2018; revised July 16, September 19, October 30, 2018; accepted October 29, 2018

Abstract   We report here on a period study and analysis of 2017 BVR$_c$I$_c$ light curves of MT Camelopardalis (GSC03737-01085). It is a solar type (T~5500 K) eclipsing binary. It was observed for six nights in December 2017 at Dark Sky Observatory (DSO) with the 0.81-m reflector. Five times of minimum light were calculated from Terrell, Gross, and Cooney's (2016, *IBVS* 6166) 2004 and 2016 observations (hereafter TGC). In addition, eleven more times were taken from the literature and six determined from the present observations. From these 15 years of observations a quadratic ephemeris was calculated:

$$JD\,Hel\,Min\,I = 2458103.66121\,d + 0.36613905 \times E - 0.000000000035 \times E2 \qquad (2)$$
$$\pm 0.00051 \quad \pm 0.00000021 \quad \pm 0.000000000015$$

A BVR$_c$I$_c$ filtered simultaneous Wilson-Devinney Program (WD) solution gives a mass ratio (0.3385 ± 0.0014), very nearly the same as TGC's (0.347 ± 0.003), and a component temperature difference of only ~140 K. As with TGC, no spot was needed in the modeling. Our modeling (beginning with BINARY MAKER 3.0 fits) was done without prior knowledge of TGC's. This shows the agreement achieved when independent analyses are done with the Wilson code. The present observations were taken 1.8 years later than the last curves by TGC, so some variation is expected.

The Roche Lobe fill-out of the binary is ~13% and the inclination is ~83.5 degrees. The system is a shallow contact W-type W UMa binary, albeit the amplitudes of the primary and secondary eclipse are very nearly identical. An eclipse duration of ~21 minutes was determined for the secondary eclipse and the light curve solution.

1. Introduction

Period studies are very important in characterizing the nature of orbital evolution of eclipsing binaries. Linear results imply that the period has been constant during the interval of observation of the binary. This gives a constant slope, O–C plot with random scatter about a horizontal line. Sudden period changes are marked by the sudden changes in slope in the plot of residuals. A quadratic result shows that the period is constantly changing—if it has a negative term, the period is decreasing. This may be due to a mass transfer so the mass ratio is approaching unity when the mass transfer is conservative. Otherwise (positive quadratic term), the mass ratio is tending to extreme values away from one. Negative quadratic terms can also be reflecting the case of angular momentum loss such as magnetic braking. Sinusoidal period changes result from light time effects due to the presence of a third body orbiting the system. In this study we find quadratic residuals. Short term quadratic changes can be a part of longer term sinusoidal curves.

2. History and observations

The variable was discovered by Nakajima *et al.* (2005) in the MISAO project as MisV1226, and identified as a W UMa

binary with a period of 0.3662 day, with a magnitude range of V = 12.93–13.54. The discovery light curve is shown in Figure 1. The binary was named MT Cam in the "78th Name List" (Kazarovets *et al.* 2006).

The system was observed at two epochs, partially in 2004 in V, and with BVI$_c$ filters on 11, 13, and 14 February 2016 by Terrell, Gross, and Cooney (2016; hereafter TGC). Their light curve analysis found component ΔT ~ 150 K, inclination = 82°,

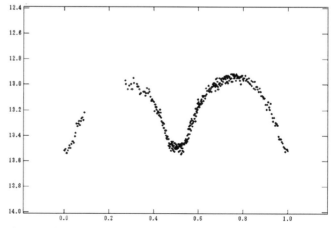

Figure 1. Light curve of Nakajima *et al.* (2005).

mass ratio, m_2/m_1 or $q = 2.88$ ($1/q = 0.35$), period $= 0.366136$ d, and $dP/dt = 1.6 \times 10^{-9}$. Their values yield a fill-out $= 7.6\%$. Their curves are displayed in Figure 2.

They assumed a $T = 5368$ K for the main component using APASS standards. Further, times of minimum light were published (Diethelm 2007, 2009, 2011, 2013; Nelson 2008).

This system was observed as a part of our professional collaborative studies of interacting binaries at Pisgah Astronomical Research Institute from data taken from Dark Sky Observatory (DSO) observations. The observations were taken by D. B. Caton. Reduction and analyses were done by Ron Samec. Our 2017 BVR_cI_c (Johnson-Cousins photometry) light curves were taken at DSO, in remote mode, with the 0.81-m DSO reflector on 5, 14, 15, 16, 17 December 2017 with a thermoelectrically cooled ($-38°$ C) $1K \times 1K$ FLI camera and Bessel BVR_cI_c filters.

Individual observations included 495 in B, 491 in V, 485 in R_c, and 491 in I_c. The probable error of a single observation was 10 mmag in B and V, 13 in R_c, and 11 mmag in I_c. The nightly C-K (Comparison-Check) star values stayed constant throughout the observing run with a precision of less than 1%. Exposure times varied from 60–100 s in B, 20–40 s in V, and 10–20 s in R_c and I_c. To produce these images, nightly images were calibrated with 25 bias frames, at least five flat frames in each filter, and ten 350-second dark frames. The BVR_cI_c observations are given in Table 1 as HJD vs Magnitude. Figures 3a and 3b show two sample B and V light curves taken 15 and 17 December 2017.

3. Finding chart

The finding chart, given here for future observers, is shown as Figure 4. The coordinates and magnitudes of the variable star, comparison star, and check star are given in Table 2.

4. Period study

Six mean times (from BVR_cI_c data averages) of minimum light were calculated from our present observations, four primary and two secondary eclipses. A least squares minimization

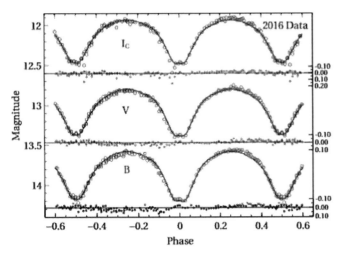

Figure 2. Observations by TGC.

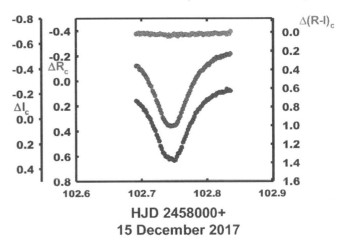

HJD 2458000+
15 December 2017

Figure 3a. Observations taken 15 December 2017. The errors for a single observation are given in section 2.

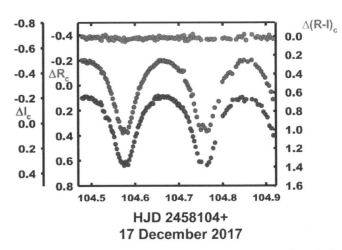

HJD 2458104+
17 December 2017

Figure 3b. Observations taken 17 December 2017. The errors for a single observation are given in section 2.

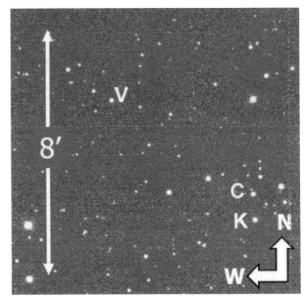

Figure 4. Finding Chart, MT Cam (V), Comparison Star (C), and Check Star (K).

method (Mikulášek *et al.* 2014) was used to determine the minima for each curve, in B,V,R$_c$, and I$_c$. These were averaged and the standard errors were determined. These are:

HJD Min I = 2458092.49374 ± 0.0002, 2458102.74600 ± 0.00007, 2458104.57686 ± 0.0002, 2458104.9434 ± 0.0029

HJD Min II = 2458103.6610 ± 0.0001, 2458104.7607 ± 0.0020.

All were weighted as 1.0 in the period study. The 2004 data (TGC) were analyzed to produce two more timings in V: HJD Min I = 2453320.7834, 2453330.6689 using the same method.

Three times of minimum light were calculated from 2016 BVI data (TGC):

HJD Min II = 2457429.7810 ±0.0005, 2457431.6126 ±0.0004, 2457432.7108 ±0.0006.

And finally, three more timings were calculated from data by Nakajima *et al.* (2005):

HJD Min I = 2452965.98833, 2452965.25301
HJD Min II = 2452975.32571

all with the same method.

These are single curves so no averaging was done and no errors are given.

Linear and quadratic ephemerides were determined from these data:

$$\text{JD Hel Min I} = 2458103.6617 + 0.366139551d \times E \quad (1)$$
$$\pm 0.0007 \pm 0.000000078$$

$$\text{JD Hel Min I} = 2458103.66121 d + 0.36613905 \times E - 0.000000000035 \times E^2 \quad (2)$$
$$\pm 0.00076 \pm 0.00000032 \pm 0.000000000022$$

The r.m.s. of the residuals for the linear and the quadratic ephemerides are given in Table 3 to for comparison. The value for the quadratic calculation is somewhat smaller. This period study covers a period of over 15 years and shows (marginally) an orbital period that is decreasing (at the 1.5 sigma level). These ephemerides were calculated by a least square O–C program. If this is a true effect, it could be due to magnetic braking that occurs as plasmas leave the system on stiff, but rotating dipole magnetic field lines. This causes angular momentum loss. This scenario is typical for overcontact binaries which eventually may coalesce due to magnetic braking, albeit in a catastrophic way producing red novae (Tylenda and Kamiński 2016). The residuals from the quadratic equation (Equation 2) are shown in Figure 5. The linear and quadratic residuals of this study are given in Table 2. The quadratic ephemeris yields a period change, $\dot{P} = 9.18 \times 10^{-8}\,\text{d/yr}$ or a mass exchange rate of

$$\frac{dM}{dt} = \frac{\dot{P} M_1 M_2}{3P (M_1 - M_2)} = \frac{-3.5 \times 10^{-8}\,M_\odot}{d} \quad (3)$$

(Qian and Zhu 2002) in a conservative scenario.

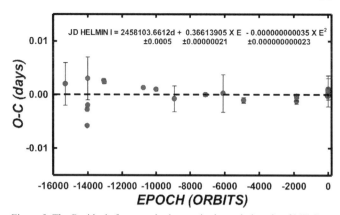

Figure 5. The Residuals from quadratic term in the period study of MT Cam. Error bars were not used as a weight for the determination of the best-fit of the quadratic ephemerides.

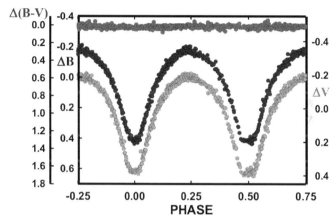

Figure 6a. B, V magnitude light curves of MT Cam phased by Equation 2.

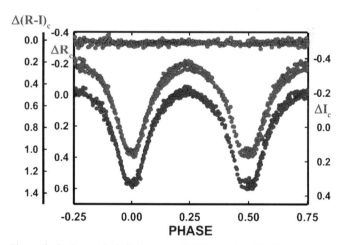

Figure 6b. R$_c$, I$_c$ magnitude light curves of MT Cam phased by Equation 2.

5. Light curve characteristics

Light curve characteristics at quadratures are shown in Figures 6a, 6b, and Table 3. The curves are of good precision, averaging about 1% photometric precision. The amplitude of the light curve varies from 0.62 to 0.55 mag in B to I. The O'Connell effect (O'Connell 1951), an indicator of spot activity, averages larger than noise level, 0.02–0.04 mag, indicating magnetic spots. The differences in minima are miniscule, averaging 0.00 mag, indicating overcontact light curves in thermal contact. A time of constant light appears to occur at minima and lasts some 21 minutes as measured by the light curve solution about phase 0.5.

6. Temperature

The 2MASS J-K for the variable corresponds to a ~G7V spectral type which yields a temperature of 5500 ± 150 K (Houdashelt *et al.* 2000; Cox 2000). This value overlaps that of TGC. This temperature was used for the light curve analysis which was done without knowledge of the TGC analysis. Fast rotating binary stars of this type are noted for having convective atmospheres, so spots are expected, but in this case were not needed in modeling.

7. Light curve solution

The B, V, R_c, I_c curves were pre-modeled with BINARY MAKER 3.0 (Bradstreet and Steelman 2002) and fits were determined and averaged from all filter bands (q~0.335, fill-out ~10%, ΔT~150 K, i~83°). The Wilson-Devinney solution (WD; Wilson and Devinney 1971; Wilson 1990, 1994; Van Hamme and Wilson 1998) was that of an overcontact eclipsing binary. The parameters were then averaged and input into a four-color simultaneous light curve calculation using WD. The solution was computed in Mode 3 (Wilson 2007) and converged to a solution. Convective parameters g=0.32, A=0.5 (Lucy 1967) were used. An eclipse duration of ~21 minutes was determined for the secondary eclipse (about phase 0.5) and the light curve solution. Thus, the binary is a W-type, W UMa binary. Since the eclipses were total, the mass ratio, q, is well determined with a fill-out of 13%. The light curve solution is given in Table 4. The Roche Lobe representation at quarter orbital phases is shown in Figures 7a, b, c, d, and the normalized fluxes overlaid by our solution of MT Cam in B, V, R_c, I_c are shown in Figures 8a and b.

8. Discussion

MT Cam is a shallow overcontact W UMa in a W-type configuration (T2 > T1). The system has a mass ratio of ~0.34, and a component temperature difference of only ~150 K. No spots were needed in the light curve modeling of the system. The Roche Lobe fill-out of the binary is ~13% with a high inclination of ~83.5° degrees. Fill-out is defined as:

$$\text{fill-out} = \frac{(\Omega_1 - \Omega_{ph})}{(\Omega_1 - \Omega_2)}, \quad (4)$$

where Ω_1 is the inner critical potential where the Roche Lobe surfaces reach contact at L_1, and Ω_2 is the outer critical potential where the surface reaches L_2.

Its spectral type indicates a surface temperature of 5500 K (Cox 2000) for the primary component, making it a solar type binary. Such a main sequence star would have a mass of 0.82 $M_\odot$ (Cox 2000) and, from the mass ratio, the secondary (from the mass ratio) would have a mass of 0.27 $M_\odot$, making it very much

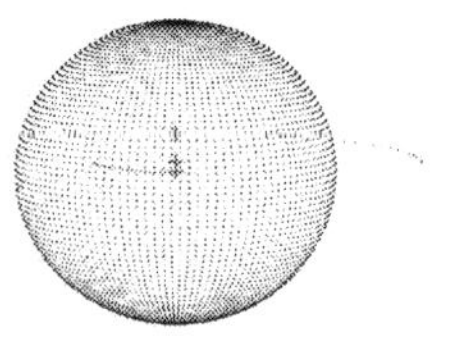

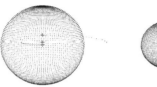

Figure 7a. MT Cam, geometrical representation at phase 0.00.

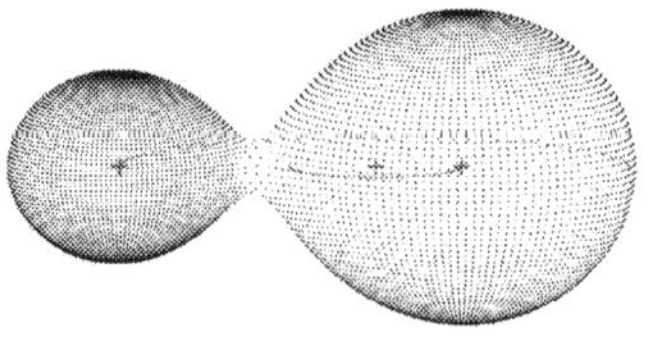

Figure 7b. MT Cam, geometrical representation at phase 0.25.

Figure 7c. MT Cam, geometrical representation at phase 0.50.

Figure 7d. MT Cam, geometrical representation at phase 0.75.

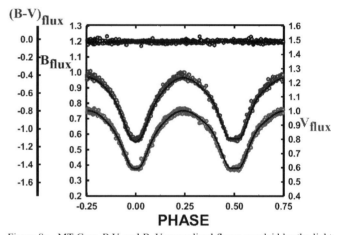

Figure 8a. MT Cam, B,V, and B–V normalized fluxes overlaid by the light curve solution.

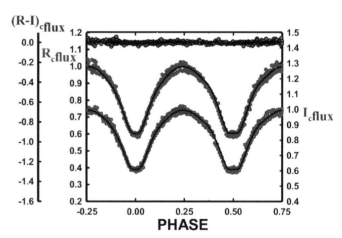

Figure 8b. MT Cam, Rc,Ic and Rc-Ic normalized fluxes overlaid by the light curve solution.

undersized. The W-type phenomena has been noted particularly on W UMa binaries that undergo interchanging depths, transit minima, and asymmetric light curves—all due to heavy spot activity (Kang *et al.* 2002). Thus, spots play a key role in this phenomena. The secondary component has a temperature of ~5645 K. Since our modeling (beginning with BINARY MAKER 3.0 fits) was done independently of TGC's, the remarkable agreement achieved shows Wilson code results are reliable. Of course, the present observations were taken 1.8 years later than the last curves by TGC, so some variation is expected.

9. Conclusions

The period study of this overcontact W UMa binary has a 15-year time duration. The period is found to be decreasing, marginally, at about the 1.5 sigma level. If the period is truly constantly decreasing, it may be due to angular momentum loss due to magnetic braking. The bifurcation in the R mag curve about phase 0.25 and I at phase 0.50 demonstrates night-to-night variation due to this solar activity. If this is the case, the system will slowly coalesce over time. In time, if this continues, one would theorize that the binary will become a rather normal, fast rotating, single G0V type field star after a small mass loss. This will probably occur following a red novae coalescence event (Tylenda and Kamiński 2016). We remind the reader that radial velocity curves are needed to obtain absolute (not relative) system parameters.

10. Acknowledgements

Dr. Samec wishes thank his collaborator, Dr. Daniel Caton, for continued observations on the 0.81-m DSO reflector.

References

Bradstreet, D. H., and Steelman, D. P, 2002, *Bull. Amer. Astron. Assoc.*, **34**, 1224.

Cox, A. N. 2000, *Allen's Astrophysical Quantities*, 4th ed., AIP Press, Springer, New York.

Diethelm, R. 2007, *Inf. Bull. Var. Stars*, No. 5781, 1.

Diethelm, R. 2009, *Inf. Bull. Var. Stars*, No. 5871, 1.

Diethelm, R. 2011, *Inf. Bull. Var. Stars*, No. 5960, 1.

Diethelm, R. 2012, *Inf. Bull. Var. Stars*, No. 6011, 1.

Diethelm, R. 2013, *Inf. Bull. Var. Stars*, No. 6063, 1.

Houdashelt, M. L., Bell, R. A., and Sweigart, A. V. 2000, *Astron. J.*, **119**, 1448.

Kang, Y. W., Oh, K-D., Kim, C-H., Hwang, C., Kim, H.-I., and Lee, W-G. 2002, *Mon. Not. Roy. Astron. Soc.*, **331**, 707.

Kazarovets, E. V., Samus, N. N., Durlevich, O. V., Kireeva, N. N., and Pastukhova, E. N. 2006, *Inf. Bull. Var. Stars* 5721, 1.

Lucy, L. B. 1967, *Z. Astrophys.*, **65**, 89.

Mikulášek, Z., Chrastina, M., Liška, J., Zejda, M., Janík, J., Zhu, L.-Y., and Qian, S.-B. 2014, *Contrib. Atron. Obs. Skalnaté Pleso*, **43**, 382.

Nakajima, K., Yoshida, S., Ohkura, N., and Kadota, K. 2005, *Inf. Bull. Var. Stars*, No. 5600, 7.

Nelson, R. H. 2008, *Inf. Bull. Var. Stars*, No. 5820, 1.

O'Connell, D. J. K. 1951, *Publ. Riverview Coll. Obs.*, **2**. 85.

Qian, S. B., and Zhu, L. Y. 2002, *Astrophys. J.*, **568**, 1004.

Skrutskie, M. F., *et al.* 2006, *Astron. J.*, **131**, 1163.

Terrell, D., Gross, J., and Cooney, W. R., Jr., 2016, *Inf. Bull. Var. Stars*, No. 6166, 1.

Tylenda, R., and Kamiński, T. 2016, *Astron. Astrophys.*, **592A**, 134.

U.S. Naval Observatory. 2012, UCAC-3 (http://www.usno. navy.mil/USNO/astrometry/optical-IR-prod/ucac).

Van Hamme, W. V., and Wilson, R. E. 1998, *Bull. Amer. Astron. Assoc.*, **30**, 1402.

Wilson, R. E. 1990, *Astrophys. J.*, **356**, 613.

Wilson, R. E. 1994, *Publ. Astron. Soc. Pacific*, **106**, 921.

Wilson, R. E. 2007, Computing binary star observables (ftp:// ftp.astro.ufl.edu/pub/wilson/lcdc2007/ebdoc2007.pdf.gz).

Wilson, R. E., and Devinney, E. J. 1971, *Astrophys. J.*, **166**, 605.

Table 1. MT Cam observations, ΔB, ΔV, ΔR$_c$, and ΔI$_c$, variable star minus comparison star.

ΔB	HJD 2458000+	ΔB	IIJD 2458000+	ΔB	HJD 2458000+	ΔB	HJD 2458000+	ΔB	HJD 2458000+
0.214	92.47346	0.308	92.69429	−0.176	101.91692	0.390	102.75267	−0.155	103.55344
0.305	92.47743	0.212	92.69713	−0.188	101.91884	0.377	102.75459	−0.161	103.55652
0.312	92.47919	0.147	92.69999	−0.184	101.92076	0.365	102.75650	−0.155	103.55961
0.349	92.48095	0.182	92.70284	−0.181	101.92268	0.321	102.75842	−0.170	103.56298
0.361	92.48271	0.133	92.70568	−0.186	101.92460	0.320	102.76033	−0.177	103.56513
0.408	92.48612	0.021	92.70853	−0.177	101.92653	0.274	102.76225	−0.160	103.56729
0.399	92.48816	0.111	92.71138	−0.201	101.92845	0.244	102.76416	−0.157	103.56944
0.412	92.49021	0.128	92.71422	−0.167	101.93037	0.198	102.76608	−0.158	103.57160
0.416	92.49226	0.011	92.71707	−0.179	101.93229	0.181	102.76799	−0.161	103.57376
0.406	92.49431	0.014	92.71992	−0.174	101.93421	0.150	102.76991	−0.165	103.57592
0.415	92.49720	−0.088	92.72277	−0.187	101.93613	0.122	102.77184	−0.154	103.57806
0.390	92.49925	−0.069	92.72561	−0.181	101.93805	0.103	102.77375	−0.156	103.58023
0.400	92.50128	−0.058	92.72846	−0.165	101.93997	0.075	102.77567	−0.147	103.58238
0.413	92.50333	−0.059	92.73131	−0.161	101.94190	0.064	102.77758	−0.163	103.58454
0.349	92.50538	−0.103	92.73414	−0.151	101.94381	0.048	102.77950	−0.143	103.58670
0.336	92.50743	−0.111	92.73700	−0.165	101.94573	0.034	102.78142	−0.144	103.58885
0.288	92.51045	−0.153	92.73985	−0.147	101.94765	−0.016	102.78526	−0.134	103.59101
0.260	92.51249	−0.147	92.74553	−0.141	101.94957	−0.015	102.78718	−0.122	103.59318
0.233	92.51452	−0.170	92.74838	−0.144	101.95148	−0.030	102.78909	−0.112	103.59532
0.194	92.51656	−0.160	92.75123	−0.108	101.95340	−0.049	102.79101	−0.119	103.59747
0.189	92.51861	0.418	101.82600	−0.105	101.95532	−0.064	102.79294	−0.100	103.59964
0.125	92.52065	0.423	101.83058	−0.102	101.95723	−0.066	102.79486	−0.094	103.60179
0.113	92.52268	0.421	101.83249	−0.078	101.95914	−0.079	102.79677	−0.086	103.60394
0.094	92.52471	0.423	101.83440	−0.091	101.96106	−0.096	102.79868	−0.083	103.60610
0.062	92.52675	0.413	101.83632	−0.065	101.96297	−0.106	102.80060	−0.052	103.60826
0.041	92.52880	0.407	101.83824	−0.077	101.96489	−0.108	102.80252	−0.055	103.61041
0.015	92.53082	0.405	101.84016	−0.072	101.96680	−0.119	102.80444	−0.045	103.61259
0.012	92.53287	0.382	101.84209	−0.046	101.96872	−0.126	102.80635	−0.025	103.61474
−0.006	92.53491	0.361	101.84400	−0.041	101.97063	−0.140	102.80827	−0.005	103.61691
−0.008	92.53694	0.342	101.84592	−0.027	101.97256	−0.145	102.81019	−0.012	103.61906
−0.013	92.53897	0.309	101.84785	−0.010	101.97447	−0.147	102.81211	0.019	103.62122
−0.067	92.54101	0.273	101.84975	0.036	101.97639	−0.149	102.81404	0.045	103.62338
−0.030	92.54303	0.244	101.85166	0.015	101.97831	−0.169	102.81596	0.068	103.62554
−0.112	92.54508	0.209	101.85357	0.069	101.98023	−0.163	102.81788	0.086	103.62772
−0.090	92.54711	0.184	101.85550	0.078	101.98215	−0.172	102.81979	0.115	103.62987
−0.082	92.54914	0.150	101.85741	−0.096	102.69307	−0.175	102.82171	0.153	103.63203
−0.111	92.55118	0.129	101.85933	−0.081	102.69500	−0.177	102.82363	0.165	103.63419
−0.086	92.55321	0.098	101.86125	−0.073	102.69692	−0.193	102.82554	0.195	103.63634
−0.125	92.55524	0.085	101.86317	−0.048	102.69884	−0.187	102.82746	0.230	103.63850
−0.117	92.55727	0.053	101.86509	−0.049	102.70076	−0.191	102.82938	0.272	103.64067
−0.135	92.55931	0.036	101.86701	−0.047	102.70269	−0.195	102.83132	0.288	103.64282
−0.142	92.56134	0.019	101.86893	−0.022	102.70461	−0.192	102.83324	0.324	103.64498
−0.141	92.56338	0.012	101.87085	−0.014	102.70654	0.416	103.48052	0.365	103.64714
−0.161	92.56542	−0.015	101.87278	0.017	102.70846	0.431	103.48335	0.405	103.64929
−0.164	92.56745	−0.022	101.87470	0.029	102.71039	0.401	103.48618	0.414	103.65145
−0.163	92.56948	−0.022	101.87662	0.036	102.71231	0.376	103.48901	0.409	103.65362
−0.169	92.57150	−0.043	101.87854	0.068	102.71423	0.344	103.49185	0.412	103.65578
−0.161	92.57354	−0.059	101.88046	0.085	102.71615	0.302	103.49494	0.419	103.65794
−0.178	92.57557	−0.072	101.88238	0.114	102.71807	0.251	103.49802	0.419	103.66010
−0.201	92.57760	−0.080	101.88430	0.128	102.72001	0.187	103.50110	0.409	103.66225
−0.171	92.57964	−0.075	101.88622	0.151	102.72193	0.164	103.50417	0.403	103.66441
−0.189	92.58168	−0.099	101.88815	0.186	102.72386	0.127	103.50725	0.439	103.66657
−0.177	92.58371	−0.103	101.89006	0.225	102.72578	0.085	103.51033	0.412	103.66872
−0.177	92.58575	−0.117	101.89198	0.235	102.72770	0.053	103.51341	0.400	103.67088
−0.150	92.58778	−0.118	101.89389	0.282	102.72963	0.025	103.51650	0.379	103.67303
−0.175	92.58981	−0.113	101.89581	0.293	102.73155	0.000	103.51958	0.356	103.67517
−0.137	92.59184	−0.121	101.89773	0.335	102.73347	−0.023	103.52266	0.326	103.67733
−0.184	92.59387	−0.151	101.89965	0.359	102.73539	−0.048	103.52572	0.284	103.67950
−0.147	92.59590	−0.145	101.90156	0.352	102.73731	−0.053	103.52882	0.264	103.68166
−0.142	92.59794	−0.133	101.90348	0.379	102.73923	−0.058	103.53189	0.224	103.68383
−0.145	92.61220	−0.168	101.90540	0.385	102.74115	−0.087	103.53497	0.183	103.68598
0.389	92.68006	−0.157	101.90731	0.406	102.74306	−0.111	103.53806	0.162	103.68813
0.397	92.68290	−0.170	101.90923	0.406	102.74498	−0.114	103.54113	0.135	103.69029
0.380	92.68575	−0.173	101.91116	0.402	102.74692	−0.124	103.54421	0.110	103.69244
0.369	92.68860	−0.158	101.91308	0.392	102.74884	−0.135	103.54729	0.095	103.69460
0.357	92.69145	−0.180	101.91500	0.395	102.75076	−0.143	103.55037	0.056	103.69677

Table continued on following pages

Table 1. MT Cam observations, ΔB, ΔV, ΔR$_c$, and ΔI$_c$, variable star minus comparison star, cont.

ΔB	HJD 2458000+	ΔB	HJD 2458000+	ΔB	HJD 2458000+	ΔB	HJD 2458000+	ΔB	HJD 2458000+
0.037	103.69892	−0.173	104.48841	0.379	104.58813	−0.121	104.68783	−0.072	104.82039
0.026	103.70107	−0.163	104.49100	0.301	104.59098	−0.118	104.69068	−0.085	104.82487
−0.006	103.70322	−0.164	104.49360	0.301	104.59098	−0.109	104.69353	−0.118	104.82934
−0.019	103.70539	−0.150	104.49619	0.295	104.59384	−0.099	104.69637	−0.159	104.83379
−0.028	103.70754	−0.151	104.49878	0.243	104.59671	−0.096	104.69922	−0.140	104.83828
−0.048	103.70969	−0.160	104.50136	0.211	104.59956	−0.079	104.70206	−0.157	104.84275
−0.051	103.71187	−0.144	104.50396	0.158	104.60242	−0.087	104.70490	−0.171	104.84723
−0.061	103.71402	−0.139	104.50655	0.160	104.60526	−0.103	104.70776	−0.148	104.85170
−0.076	103.71619	−0.131	104.50913	0.097	104.60811	−0.038	104.71062	−0.161	104.85618
−0.076	103.71833	−0.123	104.51172	0.063	104.61096	0.010	104.71347	0.145	104.86065
−0.096	103.72051	−0.105	104.51431	0.057	104.61382	−0.015	104.71633	−0.138	104.86511
−0.099	103.72267	−0.127	104.51690	0.020	104.61666	0.029	104.71919	−0.150	104.86959
−0.106	103.72483	−0.119	104.51973	0.008	104.61951	0.080	104.72203	−0.152	104.87406
−0.119	103.72700	−0.039	104.52258	−0.018	104.62236	0.057	104.72487	−0.127	104.87853
−0.120	103.72915	−0.064	104.52541	−0.030	104.62522	0.095	104.72772	−0.097	104.88299
−0.127	103.73130	−0.070	104.52827	−0.052	104.62808	0.148	104.73055	−0.079	104.88747
−0.127	103.73347	−0.027	104.53111	−0.071	104.63092	0.154	104.73340	−0.068	104.89196
−0.148	103.73597	−0.011	104.53397	−0.083	104.63377	0.217	104.73626	−0.033	104.89643
−0.154	103.73814	−0.002	104.53680	−0.082	104.63662	0.269	104.73912	−0.009	104.90090
−0.160	103.74031	0.041	104.53965	−0.101	104.63948	0.327	104.74375	0.026	104.90538
−0.154	103.74247	0.063	104.54250	−0.111	104.64232	0.309	104.74821	0.065	104.90986
−0.158	103.74464	0.092	104.54534	−0.122	104.64517	0.419	104.75269	0.132	104.91433
−0.164	103.74679	0.121	104.54819	−0.147	104.64802	0.413	104.75717	0.176	104.91882
−0.148	103.74895	0.167	104.55105	−0.141	104.65087	0.430	104.76165	0.274	104.92327
−0.167	103.75112	0.199	104.55391	−0.147	104.65371	0.436	104.76613	0.049	104.92775
−0.170	103.75327	0.240	104.55674	−0.159	104.65656	0.407	104.77060	−0.016	104.93222
−0.205	103.75544	0.269	104.55959	−0.155	104.65940	0.349	104.77508	−0.263	104.93670
−0.176	103.75759	0.327	104.56245	−0.161	104.66225	0.254	104.78015	−0.455	104.94117
−0.210	103.75976	0.363	104.56530	−0.170	104.66510	0.161	104.78463	−0.157	104.94565
−0.164	103.76180	0.398	104.56815	−0.164	104.66794	0.089	104.78911	0.411	104.95012
−0.154	103.76398	0.407	104.57102	−0.164	104.67077	0.070	104.79357	0.349	104.95460
−0.161	103.76612	0.422	104.57387	−0.154	104.67362	0.020	104.79804	−0.210	104.95908
−0.184	103.76828	0.438	104.57672	−0.154	104.67646	0.009	104.80250		
−0.151	104.48063	0.426	104.57958	−0.141	104.67930	−0.024	104.80696		
−0.169	104.48323	0.421	104.58243	−0.137	104.68214	−0.041	104.81143		
−0.160	104.48582	0.378	104.58529	−0.143	104.68500	−0.066	104.81591		

ΔV	HJD 2458000+	ΔV	HJD 2458000+	ΔV	HJD 2458000+	ΔV	HJD 2458000+	ΔV	HJD 2458000+
0.218	92.47403	0.022	92.52949	−0.194	92.58035	0.113	92.70383	0.173	101.85422
0.266	92.47801	0.001	92.53154	−0.193	92.58238	0.080	92.70668	0.163	101.85614
0.314	92.47976	−0.012	92.53358	−0.185	92.58442	0.069	92.70952	0.106	101.85807
0.331	92.48152	−0.017	92.53561	−0.195	92.58645	0.025	92.71237	0.079	101.85997
0.348	92.48327	−0.049	92.53765	−0.202	92.58848	−0.077	92.71522	0.063	101.86190
0.378	92.48683	−0.116	92.53968	−0.195	92.59051	0.014	92.71807	0.045	101.86383
0.374	92.48888	−0.058	92.54171	−0.184	92.59255	−0.006	92.72091	0.015	101.86575
0.383	92.49093	−0.060	92.54375	−0.193	92.59457	−0.074	92.72376	−0.004	101.86767
0.381	92.49297	−0.114	92.54577	0.004	92.59661	−0.048	92.72661	0.005	101.86959
0.383	92.49502	−0.133	92.54781	0.019	92.59865	−0.101	92.72946	−0.021	101.87151
0.383	92.49792	−0.140	92.54984	−0.160	92.60068	−0.311	92.73230	−0.055	101.87344
0.366	92.49995	−0.115	92.55187	−0.232	92.60883	−0.094	92.73515	−0.038	101.87536
0.366	92.50200	−0.144	92.55391	−0.081	92.61087	0.382	101.82665	−0.068	101.87728
0.374	92.50404	−0.151	92.55594	−0.091	92.61290	0.371	101.83121	−0.060	101.87920
0.329	92.50609	−0.143	92.55797	−0.123	92.61493	0.377	101.83313	−0.081	101.88112
0.283	92.50814	−0.156	92.56001	0.362	92.67537	0.378	101.83506	−0.081	101.88304
0.253	92.51116	−0.160	92.56205	0.366	92.67821	0.379	101.83698	−0.101	101.88496
0.208	92.51319	−0.168	92.56409	0.352	92.68105	0.370	101.83890	−0.122	101.88687
0.184	92.51523	−0.169	92.56612	0.382	92.68390	0.365	101.84082	−0.138	101.88879
0.172	92.51728	−0.184	92.56815	0.331	92.68675	0.338	101.84274	−0.106	101.89070
0.114	92.51932	−0.182	92.57017	0.360	92.68959	0.310	101.84466	−0.122	101.89263
0.103	92.52135	−0.176	92.57221	0.275	92.69244	0.283	101.84657	−0.154	101.89455
0.096	92.52338	−0.190	92.57424	0.226	92.69529	0.247	101.84848	−0.142	101.89647
0.114	92.52543	−0.199	92.57627	0.191	92.69814	0.233	101.85039	−0.159	101.89839
0.033	92.52746	−0.200	92.57831	0.165	92.70098	0.196	101.85231	−0.154	101.90029

Table continued on following pages

Table 1. MT Cam observations, ΔB, ΔV, ΔR$_c$, and ΔI$_c$, variable star minus comparison star, cont.

ΔV	HJD 2458000+	ΔV	HJD 2458000+	ΔV	HJD 2458000+	ΔV	HJD 2458000+	ΔV	HJD 2458000+
−0.147	101.90221	0.349	102.73797	0.177	103.50221	0.372	103.66506	−0.107	104.52356
−0.179	101.90413	0.364	102.73989	0.124	103.50529	0.398	103.66723	−0.095	104.52642
−0.182	101.90605	0.372	102.74180	0.103	103.50838	0.404	103.66938	−0.085	104.52927
−0.198	101.91182	0.370	102.74373	0.058	103.51145	0.385	103.67152	−0.077	104.53210
−0.183	101.91374	0.372	102.74565	0.019	103.51454	0.358	103.67367	−0.033	104.53495
−0.184	101.91566	0.366	102.74757	0.004	103.51762	0.334	103.67584	−0.030	104.53780
−0.204	101.91758	0.366	102.74949	−0.024	103.52070	0.303	103.67800	0.014	104.54065
−0.199	101.91950	0.370	102.75140	−0.041	103.52378	0.273	103.68015	0.039	104.54349
−0.199	101.92142	0.365	102.75332	−0.059	103.52685	0.239	103.68233	0.070	104.54635
−0.197	101.92334	0.353	102.75523	−0.070	103.52993	0.186	103.68448	0.098	104.54920
−0.194	101.92525	0.325	102.75716	−0.089	103.53301	0.164	103.68663	0.140	104.55205
−0.202	101.92717	0.300	102.75907	−0.098	103.53610	0.126	103.68879	0.179	104.55490
−0.185	101.92910	0.272	102.76099	−0.122	103.53918	0.117	103.69094	0.219	104.55774
−0.172	101.93103	0.247	102.76291	−0.129	103.54225	0.087	103.69309	0.268	104.56059
−0.188	101.93295	0.216	102.76482	−0.134	103.54533	0.055	103.69527	0.301	104.56345
−0.171	101.93487	0.184	102.76673	−0.144	103.54841	0.030	103.69742	0.332	104.56629
−0.166	101.93679	0.156	102.76866	−0.154	103.55148	0.010	103.69957	0.361	104.56915
−0.196	101.93871	0.131	102.77058	−0.163	103.55457	−0.003	103.70173	0.380	104.57201
−0.180	101.94062	0.112	102.77249	−0.165	103.55765	−0.026	103.70389	0.382	104.57487
−0.152	101.94253	0.071	102.77440	−0.172	103.56073	−0.030	103.70603	0.385	104.57773
−0.195	101.94445	0.065	102.77632	−0.168	103.56364	−0.043	103.70820	0.374	104.58058
−0.163	101.94638	0.036	102.77823	−0.166	103.56579	−0.049	103.71036	0.375	104.58343
−0.158	101.94830	0.018	102.78016	−0.182	103.56794	−0.074	103.71253	0.345	104.58628
−0.103	101.95021	0.005	102.78207	−0.185	103.57010	−0.071	103.71468	0.331	104.58913
−0.136	101.95214	−0.016	102.78400	−0.165	103.57226	−0.084	103.71683	0.276	104.59199
−0.127	101.95406	−0.042	102.78592	−0.167	103.57441	−0.080	103.71900	0.249	104.59485
−0.112	101.95596	−0.053	102.78784	−0.163	103.57657	−0.103	103.72116	0.190	104.59771
−0.099	101.95788	−0.065	102.78975	−0.164	103.57872	−0.103	103.72332	0.170	104.60055
−0.114	101.95980	−0.074	102.79167	−0.165	103.58088	−0.117	103.72549	0.121	104.60341
−0.030	101.96171	−0.085	102.79359	−0.163	103.58304	−0.116	103.72764	0.079	104.60626
−0.111	101.96362	−0.086	102.79551	−0.152	103.58520	−0.128	103.72979	0.063	104.60911
−0.079	101.96553	−0.105	102.79742	−0.147	103.58735	−0.140	103.73196	0.017	104.61196
−0.073	101.96745	−0.111	102.79933	−0.149	103.58952	−0.147	103.73413	−0.003	104.61481
−0.038	101.96937	−0.119	102.80125	−0.147	103.59167	−0.151	103.73663	−0.031	104.61766
−0.033	101.97128	−0.124	102.80317	−0.138	103.59382	−0.164	103.73880	−0.047	104.62049
−0.038	101.97320	−0.136	102.80510	−0.127	103.59598	−0.165	103.74096	−0.054	104.62335
−0.011	101.97513	−0.147	102.80702	−0.123	103.59813	−0.171	103.74313	−0.081	104.62621
0.046	101.97705	−0.151	102.80893	−0.097	103.60029	−0.167	103.74528	−0.095	104.62907
0.013	101.97897	−0.157	102.81085	−0.097	103.60460	−0.166	103.74746	−0.103	104.63192
0.018	101.98089	−0.167	102.81278	−0.084	103.60676	−0.174	103.74961	−0.117	104.63477
0.091	101.98281	−0.172	102.81469	−0.062	103.60891	−0.174	103.75177	−0.128	104.63761
0.091	101.98281	−0.179	102.81661	−0.044	103.61108	−0.177	103.75393	−0.137	104.64047
0.091	101.98281	−0.188	102.81852	−0.039	103.61324	−0.172	103.75826	−0.155	104.64333
−0.108	102.69373	−0.187	102.82044	−0.045	103.61540	−0.181	103.76042	−0.158	104.64617
−0.105	102.69565	−0.191	102.82236	−0.019	103.61756	−0.175	103.76247	−0.149	104.64901
−0.098	102.69757	−0.203	102.82429	−0.011	103.61971	−0.160	103.76462	−0.187	104.65186
−0.082	102.69949	−0.200	102.82621	0.023	103.62187	−0.208	103.76677	−0.175	104.65470
−0.070	102.70141	−0.208	102.82813	0.029	103.62404	−0.171	103.76894	−0.185	104.65754
−0.056	102.70335	−0.207	102.83005	0.057	103.62619	−0.175	103.77109	−0.187	104.66039
−0.044	102.70526	−0.197	102.83197	0.092	103.62836	−0.150	103.77324	−0.183	104.66325
−0.023	102.70719	−0.205	102.83389	0.102	103.63052	−0.183	104.48157	−0.190	104.66609
−0.006	102.70912	−0.197	102.83581	0.131	103.63269	−0.193	104.48416	−0.186	104.66893
0.011	102.71104	−0.208	102.83773	0.171	103.63484	−0.194	104.48674	−0.181	104.67178
0.037	102.71297	−0.177	102.83966	0.199	103.63701	−0.196	104.48934	−0.183	104.67462
0.053	102.71488	−0.183	102.84158	0.235	103.63916	−0.191	104.49194	−0.173	104.67745
0.071	102.71681	−0.199	102.84350	0.240	103.64132	−0.193	104.49453	−0.174	104.68030
0.090	102.71874	−0.192	102.84542	0.297	103.64348	−0.197	104.49713	−0.168	104.68313
0.123	102.72066	−0.227	102.84734	0.329	103.64563	−0.200	104.49971	−0.160	104.68599
0.152	102.72260	−0.227	102.84734	0.350	103.64779	−0.175	104.50230	−0.154	104.68883
0.175	102.72451	0.425	103.48158	0.380	103.64994	−0.176	104.50488	−0.143	104.69169
0.209	102.72644	0.391	103.48441	0.386	103.65212	−0.165	104.50747	−0.136	104.69453
0.241	102.72836	0.349	103.48724	0.399	103.65427	−0.154	104.51007	−0.132	104.69737
0.266	102.73028	0.332	103.49007	0.386	103.65644	−0.138	104.51266	−0.128	104.70021
0.289	102.73220	0.308	103.49298	0.382	103.65859	−0.153	104.51524	−0.114	104.70305
0.323	102.73412	0.257	103.49606	0.400	103.66076	−0.148	104.51789	−0.102	104.70591
0.336	102.73604	0.217	103.49913	0.393	103.66291	−0.127	104.52073	−0.107	104.70877

Table continued on following pages

Table 1. MT Cam observations, ΔB, ΔV, ΔR$_c$, and ΔI$_c$, variable star minus comparison star, cont.

ΔV	HJD 2458000+	ΔV	HJD 2458000+	ΔV	HJD 2458000+	ΔV	HJD 2458000+	ΔV	HJD 2458000+
−0.067	104.71163	0.335	104.74532	0.002	104.79515	−0.170	104.84881	−0.014	104.90247
−0.060	104.71448	0.242	104.74979	−0.091	104.79960	−0.206	104.85328	0.005	104.90695
−0.039	104.71734	0.389	104.75428	−0.042	104.80407	−0.178	104.85775	0.042	104.91143
−0.004	104.72018	0.363	104.75876	−0.078	104.80854	−0.184	104.86669	0.119	104.91590
−0.002	104.72302	0.384	104.76323	−0.103	104.81302	−0.170	104.87117	0.200	104.92038
0.049	104.72587	0.389	104.76772	−0.100	104.81750	−0.156	104.87563	0.197	104.92485
0.121	104.72871	0.343	104.77217	−0.110	104.82196	−0.144	104.88010	0.199	104.92615
0.139	104.73156	0.284	104.77665	−0.163	104.83091	−0.127	104.88457	0.256	104.95301
0.164	104.73441	0.217	104.78173	−0.168	104.83538	−0.110	104.88906		
0.219	104.73727	0.167	104.78621	−0.163	104.83986	−0.081	104.89354		
0.233	104.74011	0.136	104.79068	−0.184	104.84433	−0.057	104.89801		

ΔR$_c$	HJD 2458000+	ΔR$_c$	HJD 2458000+	ΔR$_c$	HJD 2458000+	ΔR$_c$	HJD 2458000+	ΔR$_c$	HJD 2458000+
0.158	92.47263	−0.218	92.57867	0.346	101.83349	−0.176	101.93331	0.361	102.74409
0.241	92.47661	−0.209	92.58071	0.350	101.83541	−0.224	101.93522	0.359	102.74601
0.272	92.47837	−0.201	92.58274	0.355	101.83732	−0.225	101.93715	0.359	102.74793
0.287	92.48013	−0.194	92.58478	0.346	101.83925	−0.218	101.93906	0.355	102.74985
0.313	92.48187	−0.187	92.58680	0.341	101.84118	−0.184	101.94289	0.354	102.75176
0.341	92.48515	−0.210	92.58884	0.294	101.84309	−0.173	101.94481	0.345	102.75368
0.372	92.48719	−0.195	92.59087	0.278	101.84502	−0.137	101.94673	0.326	102.75559
0.353	92.48924	−0.150	92.59290	0.252	101.84693	−0.145	101.94866	0.298	102.75751
0.355	92.49128	−0.162	92.59493	0.220	101.84884	−0.119	101.95249	0.283	102.75942
0.359	92.49333	−0.159	92.59697	0.204	101.85075	−0.156	101.95441	0.252	102.76135
0.367	92.49623	−0.125	92.59900	0.162	101.85267	−0.104	101.96015	0.225	102.76327
0.360	92.49827	0.342	92.60104	0.143	101.85458	−0.101	101.96207	0.198	102.76518
0.373	92.50031	0.363	92.60919	0.112	101.85650	−0.120	101.96398	0.198	102.76709
0.355	92.50236	0.372	92.61123	0.103	101.85841	−0.109	101.96589	0.149	102.76902
0.337	92.50441	0.327	92.61326	0.064	101.86033	−0.069	101.96781	0.121	102.77093
0.297	92.50645	0.283	92.61529	0.051	101.86225	−0.056	101.96972	0.100	102.77285
0.242	92.50947	0.252	92.61632	0.021	101.86419	−0.027	101.97164	0.077	102.77476
0.249	92.51152	0.256	92.67294	−0.001	101.86610	−0.049	101.97356	0.048	102.77668
0.204	92.51355	0.270	92.67583	−0.017	101.86994	−0.056	101.97547	0.386	103.47915
0.154	92.51559	0.120	92.67868	−0.045	101.87186	−0.053	101.97739	0.399	103.48204
0.138	92.51764	0.109	92.68153	−0.066	101.87378	0.030	101.97932	0.393	103.48487
0.117	92.51968	0.052	92.68437	−0.088	101.87764	0.023	101.98125	0.355	103.48769
0.099	92.52171	0.039	92.68722	−0.083	101.87955	0.077	101.98317	0.326	103.49053
0.074	92.52374	−0.057	92.69007	−0.107	101.88147	−0.119	102.69228	0.278	103.49350
0.031	92.52579	−0.055	92.69292	−0.086	101.88339	−0.120	102.69409	0.260	103.49658
0.013	92.52782	−0.070	92.69575	−0.125	101.88532	−0.114	102.69601	0.209	103.49966
0.004	92.52985	−0.087	92.69861	−0.120	101.88723	−0.089	102.69793	0.165	103.50274
−0.015	92.53190	−0.126	92.70146	−0.114	101.88915	−0.090	102.69985	0.115	103.50582
−0.037	92.53393	−0.123	92.70430	−0.140	101.89106	−0.077	102.70179	0.092	103.50891
−0.031	92.53597	−0.129	92.70715	−0.146	101.89298	−0.060	102.70371	0.051	103.51197
−0.071	92.53800	−0.171	92.71000	−0.149	101.89491	−0.049	102.70563	0.027	103.51507
−0.070	92.54003	−0.147	92.71285	−0.160	101.89683	−0.032	102.70755	0.008	103.51814
−0.103	92.54206	−0.198	92.71569	−0.154	101.89874	−0.015	102.70948	−0.010	103.52122
−0.137	92.54411	−0.192	92.71854	−0.166	101.90065	0.012	102.71140	−0.043	103.52430
−0.106	92.54613	0.362	92.72139	−0.197	101.90257	0.034	102.71332	−0.059	103.52738
−0.111	92.54817	0.341	92.72424	−0.162	101.90449	0.042	102.71525	−0.068	103.53046
−0.121	92.55021	0.343	92.72708	−0.175	101.90640	0.079	102.71717	−0.078	103.53354
−0.159	92.55223	0.347	92.72993	−0.200	101.90833	0.098	102.71910	−0.092	103.53663
−0.166	92.55427	0.346	92.73277	−0.203	101.91025	0.119	102.72102	−0.108	103.53969
−0.148	92.55630	0.350	92.73562	−0.218	101.91217	0.156	102.72295	−0.118	103.54277
−0.163	92.55833	0.355	92.73847	−0.204	101.91408	0.176	102.72486	−0.132	103.54585
−0.169	92.56037	0.346	92.74132	−0.205	101.91600	0.202	102.72680	−0.134	103.54893
−0.176	92.56241	0.341	92.74701	−0.178	101.91794	0.232	102.72872	−0.141	103.55201
−0.175	92.56444	0.294	92.74985	−0.209	101.91986	0.258	102.73064	−0.146	103.55509
−0.180	92.56648	0.278	92.75271	−0.176	101.92178	0.293	102.73256	−0.154	103.55818
−0.173	92.56851	0.252	92.75555	−0.233	101.92370	0.308	102.73448	−0.166	103.56188
−0.184	92.57053	0.362	101.82509	−0.218	101.92561	0.326	102.73640	−0.165	103.56404
−0.196	92.57257	0.341	101.82701	−0.180	101.92753	0.338	102.73832	−0.161	103.56620
−0.190	92.57459	0.343	101.82965	−0.173	101.92945	0.353	102.74025	−0.163	103.56835
−0.196	92.57663	0.347	101.83157	−0.232	101.93139	0.353	102.74216	−0.165	103.57051

Table continued on following pages

Table 1. MT Cam observations, ΔB, ΔV, ΔR$_c$, and ΔI$_c$, variable star minus comparison star, cont.

ΔR$_c$	HJD 2458000+	ΔR$_c$	HJD 2458000+	ΔR$_c$	HJD 2458000+	ΔR$_c$	HJD 2458000+	ΔR$_c$	HJD 2458000+
−0.164	103.57267	0.373	103.67194	0.041	103.76935	0.216	104.59539	0.076	104.73209
−0.162	103.57482	0.347	103.67408	−0.145	103.77150	0.176	104.59824	0.15	104.73494
−0.159	103.57697	0.315	103.67625	0.521	103.77365	0.134	104.60109	0.204	104.73780
−0.154	103.57914	0.291	103.67841	−0.096	103.77581	0.102	104.60394	0.254	104.74150
−0.151	103.58129	0.259	103.68056	−0.199	104.47938	0.069	104.60679	0.319	104.74597
−0.149	103.58344	0.215	103.68274	−0.191	104.48203	0.026	104.60964	0.355	104.75045
−0.143	103.58561	0.184	103.68489	0.201	104.48463	0.016	104.61250	0.318	104.75493
−0.142	103.58776	0.167	103.68704	−0.199	104.48722	−0.014	104.61534	0.342	104.75941
−0.135	103.58993	0.129	103.68920	−0.180	104.48981	−0.029	104.61819	0.367	104.76389
−0.128	103.59208	0.102	103.69135	−0.178	104.49240	−0.049	104.62104	0.359	104.76837
−0.122	103.59423	0.086	103.69350	−0.184	104.49500	−0.077	104.62390	0.317	104.77282
−0.109	103.59638	0.052	103.69568	−0.196	104.49759	−0.092	104.62676	0.248	104.77790
−0.095	103.59855	0.034	103.69783	−0.184	104.50018	−0.104	104.62960	0.18	104.78238
−0.095	103.60070	0.019	103.69998	−0.177	104.50276	−0.121	104.63245	0.239	104.78686
−0.092	103.60285	0.006	103.70214	−0.182	104.50536	−0.134	104.63530	0.129	104.79133
−0.085	103.60501	−0.018	103.70430	−0.167	104.50794	−0.14	104.63816	−0.026	104.80025
−0.077	103.60717	−0.028	103.70645	−0.162	104.51053	−0.145	104.64102	−0.056	104.80473
−0.052	103.60932	−0.036	103.70861	−0.154	104.51312	−0.159	104.64385	−0.073	104.80919
−0.057	103.61149	−0.051	103.71077	−0.153	104.51570	−0.162	104.64670	−0.096	104.81367
−0.044	103.61365	−0.067	103.71294	−0.140	104.51842	−0.178	104.64955	−0.102	104.81815
−0.020	103.61581	−0.069	103.71509	−0.134	104.52126	−0.181	104.65239	−0.09	104.82262
−0.018	103.61797	−0.079	103.71724	−0.100	104.52409	−0.197	104.65524	−0.146	104.83156
0.005	103.62012	−0.089	103.71941	−0.105	104.52695	−0.199	104.65808	−0.172	104.83603
0.027	103.62228	−0.094	103.72157	−0.094	104.52980	−0.192	104.66093	−0.197	104.84051
0.036	103.62445	−0.116	103.72373	−0.059	104.53265	−0.195	104.66378	−0.19	104.84499
0.063	103.62662	−0.108	103.72590	−0.047	104.53548	−0.198	104.66662	−0.2	104.84945
0.085	103.62877	−0.121	103.72805	−0.021	104.53834	−0.187	104.66946	−0.137	104.85392
0.108	103.63093	−0.119	103.73020	0.014	104.54119	−0.191	104.67230	−0.186	104.85840
0.141	103.63310	−0.129	103.73237	0.043	104.54402	−0.185	104.67515	−0.036	104.86287
0.172	103.63525	−0.138	103.73488	0.066	104.54688	−0.187	104.67798	−0.2	104.86735
0.203	103.63741	−0.147	103.73704	0.107	104.54973	−0.181	104.68083	−0.162	104.87628
0.234	103.63957	−0.148	103.73922	0.150	104.55259	−0.175	104.68368	−0.144	104.88075
0.263	103.64173	−0.155	103.74137	0.170	104.55544	−0.172	104.68651	−0.132	104.88522
0.301	103.64390	−0.156	103.74355	0.207	104.55827	−0.156	104.68936	−0.109	104.88971
0.324	103.64604	−0.161	103.74570	0.267	104.56112	−0.151	104.69222	−0.081	104.89419
0.384	103.65035	−0.169	103.74787	0.288	104.56398	−0.142	104.69507	−0.059	104.89865
0.387	103.65253	−0.161	103.75002	0.335	104.56683	−0.137	104.69790	−0.01	104.90313
0.391	103.65468	−0.157	103.75218	0.350	104.56970	−0.117	104.70074	0.006	104.90760
0.389	103.65685	−0.093	103.75434	0.380	104.57255	−0.111	104.70360	0.055	104.91208
0.390	103.65900	0.081	103.75650	0.351	104.57540	−0.033	104.71501	0.115	104.91656
0.399	103.66116	−0.172	103.75866	0.367	104.58111	−0.022	104.71787	0.209	104.92103
0.393	103.66332	−0.179	103.76083	0.377	104.58396	0.001	104.72071	0.235	104.92551
0.383	103.66547	−0.149	103.76288	0.332	104.58681	0.023	104.72355	−0.012	104.93892
0.393	103.66763	−0.159	103.76503	0.294	104.58966	0.039	104.72640	0.362	104.94339
0.381	103.66979	−0.156	103.76718	0.259	104.59252	0.087	104.72924	1.134	104.94787

ΔI$_c$	HJD 2458000+	ΔI$_c$	HJD 2458000+	ΔI$_c$	HJD 2458000+	ΔI$_c$	HJD 2458000+	ΔI$_c$	HJD 2458000+
0.152	92.47293	0.223	92.50978	−0.092	92.54236	−0.227	92.57693	0.302	92.68764
0.210	92.47691	0.206	92.51182	−0.153	92.54441	−0.232	92.57897	0.264	92.69049
0.231	92.47867	0.182	92.51385	−0.113	92.54643	−0.216	92.58101	0.217	92.69333
0.257	92.48043	0.134	92.51589	−0.137	92.54847	−0.235	92.58304	0.179	92.69617
0.293	92.48218	0.130	92.51794	−0.153	92.55051	−0.222	92.58508	0.070	92.69903
0.330	92.48545	0.097	92.51998	−0.177	92.55457	−0.213	92.58711	0.094	92.70187
0.331	92.48749	0.059	92.52201	−0.172	92.55660	−0.195	92.59117	0.049	92.70472
0.330	92.48954	0.031	92.52404	−0.172	92.55863	−0.182	92.59321	0.030	92.70757
0.329	92.49158	−0.007	92.52609	−0.187	92.56067	−0.190	92.59930	0.030	92.71042
0.333	92.49363	−0.031	92.52812	−0.187	92.56271	−0.228	92.61153	−0.039	92.71326
0.327	92.49653	−0.028	92.53015	−0.194	92.56474	−0.121	92.61356	−0.069	92.71611
0.301	92.49858	−0.036	92.53220	−0.187	92.56678	0.307	92.67341	−0.064	92.71896
0.337	92.50061	−0.057	92.53424	−0.201	92.56881	0.323	92.67625	−0.038	92.72180
0.319	92.50266	−0.098	92.53627	−0.199	92.57083	0.321	92.67910	−0.127	92.72465
0.281	92.50471	−0.073	92.53831	−0.213	92.57287	0.325	92.68194	−0.114	92.72750
0.253	92.50675	−0.098	92.54033	−0.214	92.57490	0.312	92.68479	−0.117	92.73035

Table continued on following pages

Table 1. MT Cam observations, ΔB, ΔV, ΔR$_c$, and ΔI$_c$, variable star minus comparison star, cont.

ΔI$_c$	HJD 2458000+	ΔI$_c$	HJD 2458000+	ΔI$_c$	HJD 2458000+	ΔI$_c$	HJD 2458000+	ΔI$_c$	HJD 2458000+
−0.200	92.73319	−0.138	101.96811	−0.164	102.80575	−0.129	103.59897	−0.208	103.75045
−0.133	92.73604	−0.101	101.97002	−0.170	102.80767	−0.116	103.60328	−0.191	103.75260
−0.151	92.73889	−0.089	101.97386	−0.183	102.80959	−0.096	103.60543	−0.200	103.75476
−0.154	92.74457	−0.016	101.97577	−0.184	102.81151	−0.093	103.60760	−0.166	103.75692
−0.207	92.75027	0.012	101.97769	−0.182	102.81344	−0.077	103.60975	−0.184	103.76119
−0.174	92.75312	0.037	101.97962	−0.199	102.81536	−0.074	103.61192	−0.191	103.76329
−0.182	92.75597	0.021	101.98154	−0.202	102.81728	−0.070	103.61408	−0.206	103.76545
0.318	101.82539	0.062	101.98347	−0.209	102.81919	−0.048	103.61624	−0.216	103.76761
0.340	101.82996	0.077	101.98317	−0.206	102.82111	−0.044	103.61839	−0.199	104.47980
0.297	101.83187	−0.139	102.69253	−0.213	102.82303	−0.018	103.62056	−0.195	104.48240
0.321	101.83378	−0.134	102.69440	−0.206	102.82494	0.004	103.62488	−0.208	104.48500
0.316	101.83570	−0.118	102.69632	−0.212	102.82686	0.058	103.62704	−0.209	104.48758
0.336	101.83763	−0.114	102.69824	−0.215	102.82879	0.073	103.62919	−0.210	104.49018
0.315	101.83955	−0.100	102.70016	−0.236	102.83072	0.088	103.63137	−0.203	104.49276
0.305	101.84148	−0.093	102.70209	−0.221	102.83264	0.116	103.63352	−0.194	104.49535
0.283	101.84339	−0.067	102.70401	−0.228	102.83456	0.145	103.63567	−0.193	104.49796
0.255	101.84531	−0.067	102.70594	−0.221	102.83648	0.276	103.64432	−0.202	104.50054
0.148	101.85296	−0.047	102.70786	−0.230	102.83840	0.301	103.64647	−0.195	104.50313
0.101	101.85488	−0.020	102.70979	−0.192	102.84032	0.321	103.64863	−0.187	104.50571
0.080	101.85680	−0.008	102.71171	−0.210	102.84224	0.346	103.65079	−0.173	104.50831
0.089	101.85871	0.005	102.71363	−0.183	102.84416	0.350	103.65295	−0.176	104.51089
0.031	101.86063	0.033	102.71555	0.353	103.47957	0.352	103.65512	−0.160	104.51348
0.017	101.86256	0.059	102.71747	0.338	103.48241	0.357	103.65727	−0.156	104.51607
−0.007	101.86448	0.080	102.71941	0.333	103.48523	0.363	103.65944	−0.152	104.51885
−0.012	101.86640	0.097	102.72133	0.326	103.48806	0.361	103.66159	−0.164	104.52169
−0.039	101.86832	0.131	102.72326	0.291	103.49088	0.352	103.66374	−0.142	104.52452
−0.042	101.87024	0.155	102.72518	0.239	103.49393	0.359	103.66591	−0.095	104.52738
−0.052	101.87216	0.180	102.72710	0.184	103.49701	0.360	103.66806	−0.089	104.53023
−0.087	101.87408	0.219	102.72903	0.149	103.50008	0.346	103.67021	−0.040	104.53307
−0.070	101.87600	0.240	102.73095	0.118	103.50316	0.337	103.67236	−0.043	104.53591
−0.109	101.87793	0.271	102.73287	0.081	103.50624	0.303	103.67451	−0.036	104.53877
−0.102	101.87985	0.280	102.73479	0.047	103.50932	0.279	103.67667	−0.022	104.54162
−0.129	101.88177	0.300	102.73671	0.015	103.51240	0.241	103.67883	0.028	104.54445
−0.132	101.88369	0.307	102.73863	−0.010	103.51548	0.219	103.68100	0.060	104.54730
−0.135	101.88561	0.305	102.74055	−0.030	103.51857	0.179	103.68316	0.077	104.55016
−0.147	101.88753	0.326	102.74246	−0.052	103.52165	0.158	103.68532	0.111	104.55302
−0.146	101.88945	0.317	102.74438	−0.090	103.52473	0.121	103.68747	0.169	104.55585
−0.169	101.89136	0.320	102.74632	−0.086	103.52780	0.101	103.68962	0.207	104.55870
−0.189	101.89328	0.329	102.74824	−0.097	103.53088	0.067	103.69177	0.248	104.56155
−0.170	101.89521	0.336	102.75016	−0.122	103.53397	0.047	103.69394	0.276	104.56441
−0.199	101.89713	0.331	102.75207	−0.129	103.53705	0.030	103.69610	0.305	104.56725
−0.155	101.89904	0.304	102.75399	−0.133	103.54012	0.005	103.69825	0.316	104.57012
−0.200	101.90095	0.279	102.75590	−0.150	103.54320	−0.001	103.70040	0.324	104.57298
−0.194	101.90479	0.274	102.75782	−0.166	103.54628	−0.033	103.70255	0.340	104.57583
−0.209	101.90670	0.252	102.75973	−0.173	103.54935	−0.026	103.70472	0.311	104.57869
−0.188	101.91054	0.229	102.76165	−0.167	103.55243	−0.052	103.70687	0.329	104.58154
−0.207	101.91246	0.189	102.76357	−0.170	103.55551	−0.074	103.70902	0.335	104.58439
−0.213	101.91631	0.166	102.76548	−0.181	103.55860	−0.073	103.71120	0.296	104.58724
−0.216	101.92208	0.143	102.76739	−0.184	103.56232	−0.085	103.71335	0.151	104.59866
−0.204	101.92399	0.040	102.77507	−0.184	103.56447	−0.100	103.71551	0.107	104.60151
−0.223	101.92591	0.029	102.77699	−0.192	103.56662	−0.106	103.71766	0.084	104.60437
−0.207	101.92783	0.010	102.77890	−0.189	103.56878	−0.108	103.71983	0.034	104.60722
−0.213	101.93169	−0.014	102.78082	−0.195	103.57094	−0.121	103.72199	0.014	104.61007
−0.204	101.93360	−0.044	102.78274	−0.192	103.57309	−0.132	103.72415	−0.015	104.61292
−0.229	101.93744	−0.051	102.78466	−0.178	103.57525	−0.135	103.72632	−0.039	104.61577
−0.223	101.94128	−0.071	102.78658	−0.188	103.57740	−0.146	103.72848	−0.072	104.61862
−0.183	101.94319	−0.075	102.78849	−0.190	103.57956	−0.144	103.73063	−0.086	104.62147
−0.182	101.94896	−0.092	102.79041	−0.175	103.58172	−0.161	103.73280	−0.098	104.62432
−0.125	101.95279	−0.100	102.79234	−0.177	103.58387	−0.160	103.73530	−0.121	104.62718
−0.159	101.95662	−0.119	102.79426	−0.170	103.58603	−0.168	103.73746	−0.124	104.63003
−0.107	101.95853	−0.128	102.79617	−0.159	103.58819	−0.168	103.73964	−0.133	104.63288
−0.137	101.96045	−0.133	102.79808	−0.161	103.59035	−0.177	103.74180	−0.149	104.63572
−0.143	101.96237	−0.150	102.80000	−0.145	103.59250	−0.178	103.74397	−0.155	104.63858
−0.139	101.96428	−0.148	102.80192	−0.147	103.59466	−0.185	103.74612	−0.152	104.64143
−0.132	101.96619	−0.158	102.80384	−0.132	103.59681	−0.165	103.74828	−0.189	104.64428

Table continued on next page

Table 1. MT Cam observations, ΔB, ΔV, ΔR$_c$, and ΔI$_c$, variable star minus comparison star, cont.

ΔI$_c$	HJD 2458000+	ΔI$_c$	HJD 2458000+	ΔI$_c$	HJD 2458000+	ΔI$_c$	HJD 2458000+	ΔI$_c$	HJD 2458000+
−0.206	104.64713	−0.175	104.69265	0.192	104.73822	−0.132	104.81880	−0.065	104.89931
−0.191	104.64997	−0.158	104.69549	0.245	104.74216	−0.181	104.82328	−0.043	104.90379
−0.215	104.65282	−0.138	104.69833	0.310	104.74663	−0.184	104.83221	0.026	104.90827
−0.191	104.65567	−0.148	104.70116	0.312	104.75111	−0.181	104.83670	0.045	104.91274
−0.203	104.65850	−0.128	104.70401	0.336	104.76454	−0.199	104.84116	0.102	104.91723
−0.208	104.66136	−0.115	104.70687	0.316	104.76902	−0.203	104.84564	0.146	104.92169
−0.211	104.66421	−0.121	104.70973	0.281	104.77348	−0.170	104.85011	0.200	104.92616
−0.202	104.66704	−0.092	104.71259	0.203	104.77856	−0.185	104.85459	0.308	104.94854
−0.205	104.66989	−0.080	104.71544	0.158	104.78304	−0.143	104.85906	0.256	104.95302
−0.203	104.67273	−0.050	104.71829	0.332	104.78752	−0.186	104.86801	0.303	104.95750
−0.210	104.67556	−0.041	104.72114	0.054	104.79199	−0.135	104.87247	0.256	104.45302
−0.199	104.67841	−0.009	104.72398	−0.023	104.79645	−0.175	104.87694	0.303	104.45750
−0.196	104.68125	0.048	104.72682	−0.053	104.80091	−0.144	104.88141	−0.835	104.43446
−0.187	104.68410	0.066	104.72966	−0.073	104.80539	−0.140	104.88589	−0.012	104.43893
−0.189	104.68694	0.110	104.73252	−0.101	104.80985	−0.133	104.89038	0.362	104.44340
−0.173	104.68979	0.146	104.73537	−0.121	104.81434	−0.086	104.89485	1.134	104.44788

Table 2. Information on the stars used in this study.

Star	Name	R.A. (2000) h m s	Dec. (2000) ° ′ ″	V	B	J–K
V	MT Cam MisV1226 GSC 3737-01085 USNO-A2.0 1425.05422897 IBVS 5600-77 2MASS J20535602-0632016 3UC167-320333	04 40 24.45	+55 25 14.4[1]	—	12.7[2]	0.450 ± 0.039[2]
C	GSC 3737-0670	04 40 56.3754	+55 22 14.215[1]	13.07[2]	—	0.42[2]
K (Check)	GSC 3737-01102 3UC291-070743	04 40 56.7551	+55 21 25.300[1]	12.65[2]	—	0.64[2]

[1] UCAC3 (*U.S. Naval Obs. 2012*). [2] 2MASS (*Skrutskie et al. 2006*).

Table 3. O–C Residuals for MT Cam.

		Epoch 2400000+	Standard[1] Error	Cycle	Linear Residual[3]	Quadratic Residual[3]	Reference
	1	52500.2629	0.0004	–15304.0	0.0008	0.0020	GCVS 5
	2	52965.2530	—	–14034.0	–0.0063	–0.0058	Nakajima *et al.* 2005
	3	52965.9883	—	–14032.0	–0.0033	–0.0028	Nakajima *et al.* 2005
	4	52975.3257	—	–14006.5	–0.0025	–0.0020	Nakajima *et al.* 2005
	5	52975.3307	0.0004	–14006.5	0.0025	0.0030	GCVS 4
	6	53320.7834	—	–13063.0	0.0026	0.0026	TGC[2]
	7	53330.6689	0.0003	–13036.0	0.0023	0.0024	TGC[2]
	8	54173.3388	—	–10734.5	0.0020	0.0013	Diethelm 2007
	9	54442.6343	0.0003	–9999.0	0.0019	0.0010	Nelson 2008
	10	54831.6560	0.0024	–8936.5	0.0003	–0.0008	Diethelm 2009
	11	55503.8891	0.0003	–7100.5	0.0012	0.0000	Diethelm 2011
	12	55875.8871	0.0035	–6084.5	0.0014	0.0003	Diethelm 2012
	13	56310.6764	0.0005	–4897.0	0.0000	–0.0010	Diethelm 2013
	14	57429.7810	0.0005	–1840.5	–0.0010	–0.0012	TGC[2]
	15	57431.6126	0.0004	–1835.5	0.0000	–0.0002	TGC[2]
	16	57432.7108	0.0006	–1832.5	–0.0003	–0.0005	TGC[2]
	17	58092.4937	0.0002	–30.5	–0.0008	–0.0002	Present observations
	18	58102.7460	0.00007	–2.5	–0.0004	0.0001	Present observations
	19	58103.6610	0.0001	0.0	–0.0008	–0.0002	Present observations
	20	58104.5769	0.0002	2.5	–0.0003	0.0003	Present observations
	21	58104.7607	0.0020	3.0	0.0005	0.0011	Present observations
	22	58104.9434	0.0029	3.5	0.0001	0.0007	Present observations
	rms				0.00202	0.00190	

1. Published or calculated errors.
2. Calculated from the light curve data given in the reference.
3. The linear and quadratic ephemerides are given in Equations 1 and 2 respectively.

Table 4. Averaged light curve characteristics of MT Cam.

Filter	Phase	Magnitude Max. I	Phase	Magnitude Max. II
	0.25		0.75	
ΔB		–0.201 ± 0.015		–0.18 ± 0.019
ΔV		–0.188 ± 0.012		–0.184 ± 0.010
ΔR		–0.191 ± 0.021		–0.195 ± 0.024
ΔI		–0.213 ± 0.016		–0.2 ± 0.015

Filter	Phase	Magnitude Min. II	Phase	Magnitude Min. I
	0.50		0.00	
ΔB		0.415 ± 0.015		0.414 ± 0.013
ΔV		0.377 ± 0.014		0.378 ± 0.007
ΔR		0.365 ± 0.021		0.363 ± 0.010
ΔI		0.334 ± 0.022		0.329 ± 0.013

Filter	Phase	Min. I – Max. I	Phase	Max. I – Max. II	Phase	Min. I – Min. II
ΔB	0.615	±0.029	–0.021	±0.034	–0.001	±0.028
ΔV	0.566	±0.020	–0.005	±0.022	0.001	±0.021
ΔR	0.554	±0.031	0.004	±0.045	–0.001	±0.031
ΔI	0.542	±0.030	–0.013	±0.032	–0.006	±0.035

Filter	Phase	Max. II –Max. I	Phase	Min. II – Max. I
ΔB	–0.015	±0.415	0.616	±0.030
ΔV	–0.012	±0.377	0.565	±0.026
ΔR	–0.021	±0.365	0.556	±0.043
ΔI	–0.016	±0.334	0.548	±0.038

Table 5. BVRI Solution Parameters, MT Cam.

Parameters	Overcontact Solution
λB, λV, λR, λI (nm)	440, 550, 640, 790
$x_{bol1,2}$, $y_{bol1,2}$	0.649 0 .649, 0.193, 0 .193
$x_{1I,2I}$, $y_{1I,2I}$	0.623, 0. 623, 0.230, 0.230
$x_{1R,2R}$, $y_{1R,2R}$	0.708, 0.708, 0. 229, 0.229
$x_{1V,2V}$, $y_{1V,2V}$	0.778, 0.778, 0. 200, 0. 200
$x_{1B,2B}$, $y_{1B,2B}$	0.847, 0.82479, 0.098, 0.098
g_1 , g_2	0.320,0.320
A_1 , A_2	0.5, 0.5
Inclination (°)	85.21 ± 0.15
T_1, T_2 (K)	5550, 5645 ± 1
$\Omega_1 = \Omega_2$ pot	2.522 ± 0.0012
$q(m_2 / m_1)$	0.3385 ± 0.0003
Fill-outs: $F_1 = F_2$ (%)	13 ± 1
$L_1 / (L_1 + L_2)_I$	0.7093 ± 0.0005
$L_1 / (L_1 + L_2)_R$	0.7064 ± 0.0006
$L_1 / (L_1 + L_2)_V$	0.7023 ± 0.0005
$L_1 / (L_1 + L_2)_B$	0.6934 ± 0.0005
JD_o (days)	2458102.74582 ± 0.00005
Period (days)	0.3661706 ± 0.0000003
r_1, r_2 (pole)	0. 452 ± 0.001, 0.276 ± 0.001
r_1, r_2 (side)	0.485 ± 0.001, 0.288 ± 0.001
r_1, r_2 (back)	0.513 ± 0.001, 0.325 ± 0.002

The First BVR$_c$I$_c$ Precision Observations and Preliminary Photometric Analysis of the Near Contact TYC 1488-693-1

Ronald G. Samec

Faculty Research Associate, Pisgah Astronomical Research Institute, 1 PARI Drive, Rosman, NC 28772; ronaldsamec@gmail.com

Daniel B. Caton

Dark Sky Observatory, Appalachian State University, Department of Physics and Astronomy, 231 Garwood Hall, 525 Rivers Street, ASU Box 32106, Boone, NC 28608-2106

Danny R. Faulkner

Johnson Observatory, 1414 Bur Oak Court, Hebron, KY 41048

Robert Hill

Bob Jones University, 1700 Wade Hampton Boulevard, Greenville, SC 29614

Received April 24, 2018; revised June 26, July 10, August 6, 2018; accepted August 27, 2018

Abstract   TYC 1488-693-1 is an ~F2 type (T~6750 K) eclipsing binary. It was observed in April and May 2015 at Dark Sky Observatory in North Carolina with the 0.81-m reflector of Appalachian State University. Six times of minimum light were determined from our present observations, which include two primary eclipses and four secondary eclipses. In addition, six observations at minima were determined from archived NSVS Data. Improved linear and quadratic ephemerides were calculated from these times of minimum light which gave a possible period change of dP/dt = –5.2 (1.5) × 10^{-6} d/yr. The period decrease may indicate that the binary is undergoing magnetic braking and is approaching a contact configuration due to the angular momentum loss. A BVR$_c$I$_c$ simultaneous Wilson-Devinney (WD) Program solution indicates that the system has a mass ratio (q = M$_2$/M$_1$) of ~0.58 (our solutions taken from q = 0.3 to 1.2 also indicate this is the value with the lowest sum of square residual), and a component temperature difference of ~2350 K. The large ΔT in the components verifies that the binary is not in contact. A BINARY MAKER fitted hot spot was maintained in the WD Synthetic Light Curve Computations. It remained on the larger component at the equator on the correct (following) side for a stream spot directed from the secondary component (as dictated by the Coriolis effect). This could indicate that the components are near filling their respective Roche Lobes. The fill-outs are nearly identical, 96% for the primary component and 95% for the secondary component. The inclination is ~79°, which is not enough for the system to undergo a total eclipse. Caution is given for taking this solution as the definitive one.

1. Introduction

We expect solar type contact binaries to have begun their evolution as pre-contact, detached binaries (Qian, Zhu, and Boonruksar 2006; Samec *et al.* 2015; Guinan and Bradstreet 1988). We are finding dwarf F though K type binaries in this configuration (Samec *et al.* 2017a; Samec *et al.* 2012). Magnetic braking is the probable physical mechanism responsible. TYC 1488-693-1 is such a binary. It apparently is near the detached-contact boundary of this evolution. We present a photometric analysis of this binary in the following sections.

2. History and observations

TYC 1488-693-1 (NSVS 10541123) is listed in the All Automated Sky Survey (Pojmański *et al.* 2013). Light curve data (Figure 1) are given at the NSVS website (Los Alamos Natl. Lab. 2017), which is a SkyDOT database for objects in time-domain. The binary is in the constellation of Cancer. The All Sky Automated Survey-3 (Pojmański *et al.* 2013) categorizes it as a semi-detached eclipsing binary (ESD) type with an amplitude of 0.71 V, and a period of 0.59549 d, ID = 145957+1938.6.

VSX (Watson *et al.* 2014) gives magnitude range of V = 11.87 – 12.7 and characterizes it as an EA (Algol) type.

This system was observed as a part of our studies of interacting binaries from Shaw's Near Contact Binaries (e.g., Caton *et al.* 2018; Samec *et al.* 2016, 2017b) with data taken from Dark Sky Observatory observations (DSO; Appalachian State Univ. 2018). The curves shown are from the NSVS Catalog entry Object 10541123.

The new GAIA DR2 (Bailer-Jones 2015) results give a distance of 760 ±40 pc.

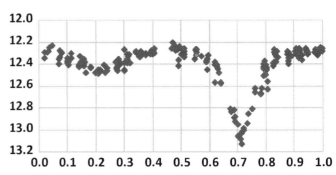

Figure 1. Data from the ASAS NSVS catalog entry object 10541123 (Los Alamos Natl. Lab. 2017).

Table 1. Information on the stars used in this study.

Star	Name	R.A. (2000) h m s	Dec. (2000) ° ' "	V	J–K
V	NSVS 10541123 TYC 1488-693-1 ASAS 45957+1938.6 CRTS J145957.0+19383 2MASS J14595711+1938393	14 59 57.0904	+19 38 39.458	11.75[1]	0.18 ± 0.04[1]
C	TYC 1488-723-1	14 59 44.2342	+19 36 50.878	11.24[2]	0.34 ± 0.04[2]
K (Check)	TYC 1488-641-1 BD+20 3050	14 59 23.5159	+19 41 48.135[2]	10.51[2]	0.54 ± 0.08[2]

[1] 2MASS (Skrutskie et al. 2006). [2] ICRS (U. S. Naval Obs. 2018).

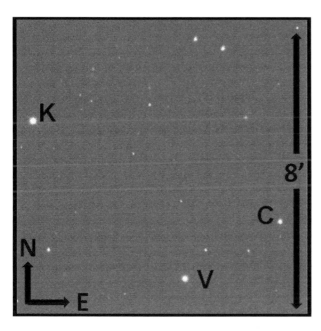

Figure 2. Finder chart showing V image of TYC 1488-693-1 (the variable V, the comparison star C, and the check star K).

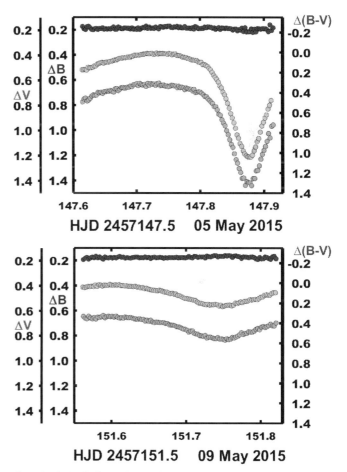

Figure 3a (top), 3b (bottom). Sample TYC 1488-693-1 observations of B, V, and B–V color curves on the nights of 5 and 9 May 2015.

Our BVR$_c$I$_c$ light curves were taken with the DSO 0.81-m f/8 R-C reflector at Philips Gap, North Carolina, on 1, 5, 24 April and 5, 6, 8, 9, 10 May 2015 with a thermoelectrically cooled (–40°C) 2K × 2K Apogee Alta by D. Caton, R. Samec, and D. Faulkner with BVR$_c$I$_c$ filters. Reductions were done with AIP4WIN V2. Individual observations included 751 in B, 566 in V, 767 in R$_c$, and 741 in I$_c$. The probable error of a single observation was 6 mmag B, 10 mmag in V, 6 mmag in R$_c$, and 7 mmag in I$_c$. The nightly C–K values stayed constant throughout the observing run with a precision of 0.1–1%. Exposure times varied from 100 s in B to 15 s in V, R$_c$, and I$_c$. Nightly Images were calibrated with 25 bias frames, at least five flat frames in each filter, and ten 300-s dark frames.

3. Stellar identifications and finding chart

The coordinates and magnitudes of the variable star, comparison star, and check star are given in Table 1. The finding chart is shown as Figure 2.

Figures 3a and 3b show sample observations of B, V, and B–V color curves on the night of 5 and 9 May 2015. Our observations are given in Table 2, in delta magnitudes, ΔB, ΔV, ΔR$_c$, and ΔI$_c$, in the sense of variable minus comparison star.

4. Period study

Six mean times of minimum light were calculated, two primary and four secondary eclipses, from our present B,V,R$_c$,I$_c$ observations:

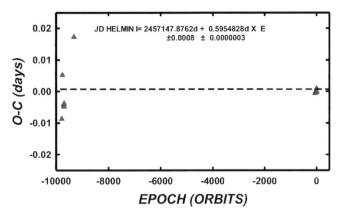

Figure 4a. Linear residuals of equation 1, TYC 1488-693-1; the residual r.m.s. = 0.0215.

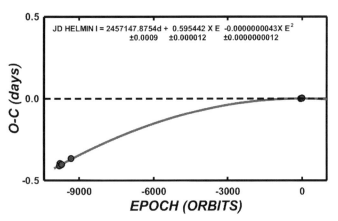

Figure 4b. Residuals of equation 2 vs. quadratic term, TYC 1488-693-1; the residual r.m.s. = 0.0120.

HJD I = 2457113.93303 ± 0.00020, 2457147.87606 ± 0.000010
HJD II = 2457117.80391 ± 0.00049, 2457136.85995 ± 0.00068,
  2457148.77040 ± 0.00037, 2457151.7468 ± 0.0002

Six times of low light were taken from an earlier light curve phased from data (NSVS 10541123; Los Alamos Natl. Lab. 2017) in the Northern Sky Variability Survey. Figure 1 was used to get times of minima within ± 0.01 phase unit.

A linear ephemeris and quadratic ephemerides were determined from these data and are given next. The given errors are standard errors.

$$JDHelMinI = 2457147.8762 + 0.5954828 \text{ d} \times E \qquad (1)$$
$$\pm 0.0008 \quad \pm 0.0000003$$

$$JDHelMinI = 2457147.87539 \text{ d} + 0.595442 \times E - 0.0000000043 \times E^2 \quad (2)$$
$$\pm 0.00089 \quad \pm 0.000012 \quad \pm 0.0000000012$$

The period study covers a period of some 16 years and shows a period that is decreasing. The problem with this fit is the large gap of time (nearly 9 years) between the last of the Skydot points and the first point of the present observations. So this result cannot be taken as definitive. The rate of orbital period change, $dP/dt = -5.2\,(1.5) \times 10^{-6}$ d/yr, is high for detached systems, according to my unpublished study of some 200 solar type binaries which appeard at the 2018 IAU GA. The residuals from the linear and quadratic period study are

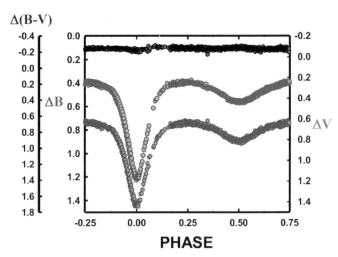

Figure 5a. ΔB, ΔV light and Δ(B–V) color curves folded using Equation (1) of TYC 1488-693-1, delta mag vs. phase.

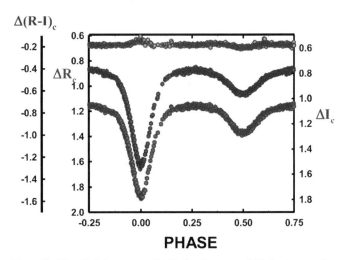

Figure 5b. ΔRc, ΔI_c light curves and Δ(R–I)$_c$ color curves folded using equation (1) of TYC 1488-693-1, delta mag vs phase.

given in Table 3. These residuals are plotted in Figures 4a and 4b, respectively.

5. Light curve characteristics

The phased B,V and R_c,I_c light curves folded using Equation (1), delta mag vs. phase, are shown in Figures 5a and 5b, respectively. Light curve characteristics are tabulated by quadrature (averaged magnitudes about Phase 0.0, 0.25, 0.50, and 0.75) in Table 4. As noted in the table, averaged data about phase 0.0 (primary eclipse) are denoted as "Min I", phase 0.5 (secondary eclipse) as "Min II", phase 0.25 as "Max I", and phase 0.25 as "Max II". The folded light curves are of good precision, averaging somewhat better than 1% photometric precision. The primary amplitude of the light curve varies from 0.81–0.71 mag in B to I_c, indicating a substantial inclination. The secondary amplitude varies from 0.17 to 0.22 mag B to I_c, respectively. The O'Connell effect, an indicator of spot activity visible during Max I and Max II, is nearly 0.0 mag for all filters. This, however, does not preclude the possibility of other spot(s). In this case we found that a hot spot located on the primary star facing its binary partner was necessary to achieve the best fit.

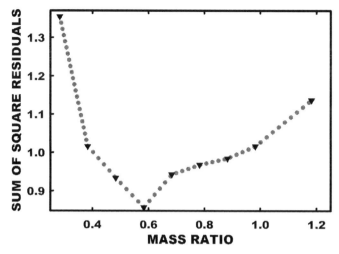

Figure 6. Various solutions at different fixed mass ratios (0.3 to 1.2) vs. the sum of square residuals of each, indicated by inverted triangles. The lowest sum of sum of square residuals occurred at mass ratio was at q~0.6. This chart may be helpful in limiting a mass ratio to model when radial velocities become available.

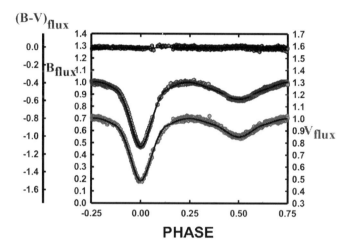

Figure 7a. Solution overlaying B, V normalized flux light curves for TYC 1488-693-1.

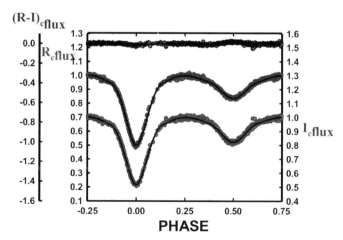

Figure 7b. Solution overlaying R$_c$, I$_c$ normalized flux light curves for TYC 1488-693-1.

Figure 8a. Roche lobe stellar surface at phase 0.00 of TYC 1488-693-1.

Figure 8b. Roche lobe stellar surface at phase 0.25 of TYC 1488-693-1.

  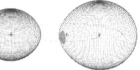

Figure 8c. Roche lobe stellar surface at phase 0.50 of TYC 1488-693-1.

Figure 8d. Roche lobe stellar surface at phase 0.75 of TYC 1488-693-1.

The absence of the spot affects the shoulders of the curves, especially the ingress of the primary eclipse.

The differences in minima are large, 0.50–0.65 mag in I$_c$ to B, respectively, indicating noncontact components. The fact is easily seen in the R$_c$–I$_c$ color curves rising at phase zero and dipping at the secondary eclipse. This is a sign of a near contact system, Algol type.

6. Temperature and light curve solution

The *Tycho* photometry (Høg *et al.* 2000) gives a B–V = 0.409 ± 0.153 (T~6600 K) and the 2MASS (*Skrutskie et al. 2006*) J–K = 0.18 ± 0.04 (T~6885 K) for the binary. This corresponds to ~F2V ± 2, or a temperature of about 6750 ± 150 K. Fast rotating binary stars of this type are noted for still having magnetic activity (Samec *et al.* 2007), but this phenomena should not dominate. This binary may be becoming an early type W UMa binary.

The B, V, R$_c$, and I$_c$ curves were pre-modeled with BINARY MAKER 3.0 (Bradstreet and Steelman 2002) and fits were determined in all filter bands. The result of the best fit was that of a detached eclipsing binary with both components underfilling filling their critical Roche lobes. The parameters were then averaged and input into a four-color simultaneous light curve calculation using the Wilson-Devinney Program (Wilson and Devinney 1971; Wilson 1990, 1994; Van Hamme and Wilson 1998). The solution was computed in Mode 2 so that each potential was allowed to adjust and thus the configuration was completely determined by the calculation. The solution converged in a detached configuration with a q~0.58. Convective parameters, g = 0.32, A = 0.5 were used. Since the eclipses were not total, a number of solutions were generated with fixed mass ratios (q). These were iterated with spot parameters.

The sum of square residuals was tabulated with a q-value from 0.3 to 1.2. As in many cases, the original solution was found to have the lowest sum of square residuals with a q = 0.58. The mass ratio vs. residual plot is shown graphically in Figure 6. The best solution is given in Table 5. The normalized curves overlain by our light curve solutions are shown as Figure 7a, and 7b. A geometrical (Roche-lobe) representation of the system is given in Figure 8 a,b,c,d at light curve quadrature's so that the reader may see the placement of the spot and the relative size of the stars as compared to the orbit. We note here that a mass ratio search is generally not sufficient to determine the mass ratio. But it is attempted here to offer some constraint to the system's characteristics. Along this line, I note that an anonymous referee stated that a number of well fit unspotted solutions were found with mass ratios between 0.4 and 0.7.

A precision radial velocity curve is needed to find the true mass ratio.

7. Discussion

TYC 1488-693-1 is found to be a detached, near contact binary. Both Roche lobes are over 90% filled, potential-wise. The photometric spectral type indicates a surface temperature of 6750 K for the primary component. The secondary component has a temperature of $\sim$4570 K (K4V). Our mass ratio is $\sim$0.6, with an amplitude of 0.8–0.7 mag in B to I_c, respectively. The spot on the primary component is at the physical position that a stream spot would be expected, so it is possible that a weak plasma stream is being emitted from the secondary component with the primary component as the gainer. The inclination is 79°, which allows only 2% of the light of the system to be contributed by the secondary component at phase 0.5. The iterated hot spot region has a 9° radius and a mean T-factor of 1.05 (T$\sim$7100 K). The mass ratio and the component temperatures indicate that the secondary is somewhat oversized so that interactions may have occurred in the past.

8. Conclusions

The period study of this near contact binary has a 16-year duration. But it is sparse. The orbital period was found to be decreasing. This decrease is not unusual for a typical W UMa system undergoing magnetic braking. The composite (8 nights) light curve's higher noise level as compared to that of single nights' curves may indicate solar type activity. If this is the case, the system will come into contact and then slowly coalesce over time as it loses angular momentum due to ion winds moving radially outward on stiff magnetic field lines rotating with the binary (out to the Alfvén radius). We expect the system is tending to become a W UMa contact binary and, ultimately, will become a rather normal, fast rotating, single $\sim$A2V type field star following a red novae coalescence event when both components merge (Tylenda and Kamiński 2015).

Radial velocity curves are needed to obtain absolute (not relative) system parameters and a firm mass ratio.

9. Acknowledgements

Dr. Samec wishes to thank Dark Sky Observatory for continued use of the 32-inch observing facilities of Appalachian State University. This work has made use of data from the European Space Agency (ESA) mission Gaia (https://www.cosmos.esa.int/gaia), processed by the Gaia Data Processing and Analysis Consortium (DPAC, https://www.cosmos.esa.int/web/gaia/dpac/consortium). Funding for the DPAC has been provided by national institutions, in particular the institutions participating in the Gaia Multilateral Agreement.

References

Appalachian State University, Boone, North Carolina. 2018, observations from the Dark Sky Observatory (DOS; https://dso.appstate.edu/).

Bailer-Jones, C. A. L. 2015, *Publ. Astron. Soc. Pacific*, **127**, 994.

Bradstreet, D. H., and Steelman, D. P. 2002, *Bull. Amer. Astron. Soc.*, **34**, 1224.

Caton, D. B., Samec, R. G., and Faulkner, D. R. 2018, Amer. Astron. Soc. Meeting #231, id. 244.03.

Guinan, E. F., and Bradstreet, D. H. 1988. in *Formation and Evolution of Low Mass Stars*, eds. A. K. Dupree and M. T. V. T. Lago, NATO Adv. Sci. Inst. (ASI) Ser. C, 241. Kluwer, Dordrecht, Netherlands, 345.

Høg, E., *et al.* 2000, *Astron. Astrophys.*, **355**, L27.

Los Alamos National Laboratory. 2017, SkyDOT Northern Sky Variability Survey database (http://skydot.lanl.gov/star.php?num=10083189&mask=32004).

Pojmański, G., Szczygiel, D., and Pilecki, B. 2013, The All-Sky Automated Survey Catalogues (ASAS3; http://www.astrouw.edu.pl/asas/?page=catalogues).

Qian, S.-B., Zhu, L.-Y., and Boonruksar, S. 2006, *New Astron.*, **11**, 503.

Samec, R. G., Clark, J. D., Van Hamme, W., and Faulkner, D. R. 2015, *Astron. J.*, **149**, 48.

Samec, R. G., Dignan, J. G., Smith, P. M., Rehn, T., Oliver, B. M., Faulkner, D. R., and Van Hamme, W. 2012, *Inf. Bull. Var. Stars*, No. 6035, 1.

Samec, R. G., Jones, S. M., Scott, T., Branning, J., Miller, J., Chamberlain, H., Faulkner D. R., and Hawkins, N. C. 2007, in *Convection in Astrophysics*, eds. F. Kupka, I. W. Roxburgh, K. L. Chan, Proc. IAU Symp. 239, Cambridge University Press, Cambridge, 505.

Samec, R. G., Nyaude, R., Caton, D., and Van Hamme, W. 2016, *Astron. J.*, **152**, 199.

Samec, R. G., Olsen, A., Caton, D., and Faulkner, D. R. 2017a, *J. Amer. Assoc. Var. Star Obs.*, **45**, 148.

Samec, R. G., Olsen, A., Caton, D. B., Faulkner, D. R., and Hill, R. L. 2017b, *J. Amer. Assoc. Var. Star Obs.*, **45**, 173.

Skrutskie, M. F., *et al.* 2006, *Astron. J.*, **131**, 1163.

Tylenda, R., and Kamiński, T. 2016, *Astron. Astrophys.*, **592A**, 134.

U. S. Naval Observatory. 2018, International Celestial Reference System (ICRS; http://aa.usno.navy.mil/faq/docs/ICRS_doc.php).

Van Hamme, W. V., and Wilson, R. E. 1998, *Bull. Amer. Astron. Soc.*, **30**, 1402.

Watson, C., Henden, A. A., and Price, C. A. 2014, AAVSO International Variable Star Index VSX (Watson+, 2006–2014; http://www.aavso.org/vsx).

Wilson, R. E. 1990, *Astrophys. J.*, **356**, 613.

Wilson, R. E. 1994, *Publ. Astron. Soc. Pacific*, **106**, 921.

Wilson, R. E., and Devinney, E. J. 1971, *Astrophys. J.*, **166**, 605.

Table 2. TYC 1488-693-1 observations, ΔB, ΔV, ΔR$_c$, and ΔI$_c$, variable star minus comparison star.

ΔB	HJD 2457000+	ΔB	HJD 2457000+	ΔB	HJD 2457000+	ΔB	HJD 2457000+	ΔB	HJD 2457000+
0.436	113.7145	0.499	113.8712	0.552	117.8033	0.417	136.6614	0.515	136.8202
0.437	113.7169	0.513	113.8736	0.551	117.8056	0.418	136.6638	0.522	136.8227
0.437	113.7193	0.525	113.8760	0.553	117.8077	0.417	136.6663	0.525	136.8251
0.434	113.7218	0.543	113.8785	0.557	117.8099	0.414	136.6687	0.526	136.8276
0.435	113.7242	0.561	113.8809	0.555	117.8121	0.408	136.6712	0.538	136.8300
0.431	113.7266	0.582	113.8833	0.556	117.8143	0.408	136.6736	0.533	136.8324
0.432	113.7290	0.600	113.8857	0.551	117.8165	0.407	136.6760	0.542	136.8349
0.426	113.7314	0.625	113.8881	0.553	117.8186	0.405	136.6785	0.542	136.8373
0.428	113.7338	0.645	113.8905	0.547	117.8208	0.404	136.6809	0.548	136.8398
0.422	113.7362	0.671	113.8929	0.538	117.8230	0.403	136.6834	0.542	136.8422
0.421	113.7387	0.703	113.8953	0.540	117.8252	0.401	136.6858	0.550	136.8446
0.413	113.7411	0.732	113.8977	0.539	117.8274	0.393	136.6883	0.552	136.8471
0.416	113.7435	0.767	113.9001	0.537	117.8296	0.393	136.6907	0.557	136.8495
0.412	113.7459	0.806	113.9025	0.537	117.8317	0.393	136.6932	0.560	136.8520
0.415	113.7483	0.844	113.9050	0.535	117.8339	0.384	136.6956	0.562	136.8544
0.405	113.7508	0.886	113.9074	0.531	117.8361	0.386	136.6981	0.567	136.8568
0.400	113.7532	0.919	113.9098	0.531	117.8383	0.387	136.7005	0.560	136.8593
0.398	113.7556	0.962	113.9122	0.522	117.8405	0.389	136.7030	0.564	136.8617
0.399	113.7580	0.999	113.9146	0.516	117.8426	0.391	136.7055	0.561	136.8642
0.394	113.7604	1.041	113.9170	0.512	117.8448	0.390	136.7079	0.559	136.8666
0.391	113.7628	1.084	113.9194	0.502	117.8470	0.388	136.7104	0.559	136.8690
0.387	113.7653	1.118	113.9218	0.496	117.8492	0.393	136.7128	0.556	136.8715
0.387	113.7677	1.144	113.9242	0.493	117.8514	0.392	136.7153	0.545	136.8739
0.390	113.7701	1.173	113.9267	0.487	117.8535	0.395	136.7177	0.548	136.8764
0.387	113.7725	1.193	113.9291	0.487	117.8557	0.396	136.7202	0.543	136.8788
0.389	113.7749	1.202	113.9315	0.480	117.8579	0.392	136.7226	0.541	136.8812
0.388	113.7773	1.202	113.9339	0.487	117.8601	0.396	136.7250	0.544	136.8837
0.387	113.7797	1.200	113.9363	0.475	117.8623	0.401	136.7275	0.540	136.8861
0.386	113.7821	1.183	113.9387	0.478	117.8644	0.400	136.7299	0.533	136.8886
0.384	113.7845	1.155	113.9411	0.475	117.8666	0.402	136.7324	0.540	136.8910
0.385	113.7869	0.450	117.7269	0.475	117.8688	0.403	136.7348	0.530	136.8934
0.383	113.7893	0.452	117.7291	0.469	117.8710	0.403	136.7373	0.527	136.8959
0.382	113.7917	0.454	117.7313	0.472	117.8732	0.406	136.7397	0.523	136.8983
0.382	113.7942	0.460	117.7335	0.458	117.8754	0.404	136.7422	0.524	136.9008
0.386	113.7966	0.463	117.7357	0.461	117.8775	0.406	136.7446	0.515	136.9032
0.387	113.7990	0.466	117.7379	0.452	117.8797	0.409	136.7471	0.512	136.9057
0.385	113.8014	0.475	117.7400	0.455	117.8819	0.414	136.7495	0.499	136.9081
0.388	113.8038	0.475	117.7422	0.451	117.8841	0.418	136.7520	0.499	136.9105
0.389	113.8062	0.474	117.7444	0.452	117.8863	0.423	136.7544	0.503	136.9130
0.391	113.8086	0.483	117.7466	0.449	117.8884	0.431	136.7569	0.492	136.9154
0.390	113.8110	0.487	117.7488	0.445	117.8906	0.429	136.7593	0.490	136.9179
0.396	113.8134	0.491	117.7509	0.441	117.8928	0.432	136.7617	0.521	147.6143
0.393	113.8158	0.491	117.7531	0.443	117.8950	0.435	136.7641	0.520	147.6167
0.394	113.8182	0.489	117.7553	0.442	117.8972	0.441	136.7666	0.519	147.6192
0.400	113.8206	0.494	117.7575	0.432	117.8993	0.437	136.7690	0.523	147.6216
0.399	113.8231	0.498	117.7597	0.432	117.9015	0.442	136.7715	0.501	147.6240
0.403	113.8255	0.504	117.7618	0.426	117.9037	0.438	136.7739	0.505	147.6265
0.409	113.8279	0.509	117.7640	0.425	117.9059	0.443	136.7764	0.504	147.6289
0.410	113.8303	0.518	117.7662	0.423	117.9081	0.446	136.7788	0.500	147.6314
0.416	113.8327	0.521	117.7684	0.416	117.9103	0.445	136.7812	0.491	147.6338
0.416	113.8351	0.529	117.7706	0.412	117.9124	0.450	136.7837	0.490	147.6362
0.418	113.8375	0.536	117.7728	0.409	117.9146	0.455	136.7861	0.476	147.6387
0.421	113.8399	0.534	117.7749	0.412	117.9168	0.459	136.7885	0.470	147.6411
0.425	113.8423	0.544	117.7771	0.412	117.9190	0.460	136.7910	0.465	147.6435
0.431	113.8447	0.542	117.7793	0.405	117.9211	0.465	136.7934	0.463	147.6460
0.432	113.8471	0.542	117.7815	0.409	117.9233	0.466	136.7959	0.468	147.6484
0.436	113.8495	0.545	117.7837	0.409	117.9255	0.476	136.7983	0.463	147.6509
0.442	113.8519	0.547	117.7858	0.413	117.9277	0.482	136.8007	0.459	147.6533
0.444	113.8544	0.546	117.7880	0.411	117.9299	0.486	136.8032	0.460	147.6557
0.446	113.8568	0.551	117.7902	0.412	117.9320	0.489	136.8056	0.455	147.6582
0.453	113.8592	0.549	117.7924	0.413	117.9342	0.493	136.8081	0.453	147.6606
0.457	113.8616	0.548	117.7946	0.410	117.9364	0.497	136.8105	0.448	147.6631
0.467	113.8640	0.554	117.7968	0.410	117.9386	0.503	136.8129	0.443	147.6655
0.475	113.8664	0.553	117.7990	0.399	117.9408	0.500	136.8154	0.444	147.6679
0.485	113.8688	0.550	117.8011	0.422	136.6590	0.506	136.8178	0.437	147.6704

Table continued on following pages

Table 2. TYC 1488-693-1 observations, ΔB, ΔV, ΔR$_c$, and ΔI$_c$, variable star minus comparison star, cont.

ΔB	HJD 2457000+	ΔB	HJD 2457000+	ΔB	HJD 2457000+	ΔB	HJD 2457000+	ΔB	HJD 2457000+
0.430	147.6728	0.648	147.8311	0.517	148.8130	0.438	150.6635	0.409	151.6275
0.429	147.6753	0.664	147.8336	0.510	148.8154	0.435	150.6660	0.410	151.6299
0.427	147.6777	0.697	147.8360	0.507	148.8179	0.431	150.6684	0.408	151.6323
0.424	147.6801	0.718	147.8385	0.501	148.8203	0.435	150.6709	0.413	151.6348
0.425	147.6826	0.750	147.8409	0.496	148.8227	0.427	150.6733	0.415	151.6372
0.421	147.6850	0.786	147.8433	0.497	148.8252	0.427	150.6758	0.420	151.6397
0.419	147.6874	0.825	147.8458	0.494	148.8276	0.426	150.6782	0.418	151.6421
0.421	147.6899	0.868	147.8482	0.487	148.8300	0.419	150.6806	0.426	151.6445
0.413	147.6923	0.903	147.8506	0.488	148.8325	0.419	150.6830	0.424	151.6470
0.412	147.6947	0.943	147.8531	0.487	148.8349	0.413	150.6855	0.430	151.6494
0.405	147.6972	0.986	147.8555	0.484	148.8373	0.410	150.6879	0.434	151.6519
0.405	147.6996	1.039	147.8580	0.480	148.8398	0.406	150.6904	0.437	151.6543
0.403	147.7021	1.065	147.8604	0.476	148.8422	0.408	150.6928	0.438	151.6568
0.404	147.7045	1.108	147.8628	0.475	148.8446	0.407	150.6952	0.441	151.6592
0.400	147.7069	1.130	147.8653	0.467	148.8470	0.404	150.6977	0.443	151.6616
0.398	147.7094	1.165	147.8677	0.460	148.8495	0.404	150.7001	0.439	151.6641
0.396	147.7118	1.190	147.8701	0.459	148.8519	0.406	150.7025	0.442	151.6665
0.393	147.7142	1.204	147.8726	0.457	148.8543	0.411	150.7050	0.449	151.6689
0.390	147.7167	1.213	147.8750	0.455	148.8568	0.408	150.7074	0.454	151.6714
0.393	147.7191	1.214	147.8774	0.455	148.8592	0.408	150.7098	0.453	151.6738
0.392	147.7215	1.215	147.8799	0.448	148.8616	0.410	150.7122	0.460	151.6763
0.390	147.7240	1.200	147.8823	0.448	148.8641	0.416	150.7147	0.466	151.6787
0.398	147.7264	1.166	147.8848	0.441	148.8665	0.404	150.7171	0.464	151.6812
0.395	147.7289	1.135	147.8872	0.440	148.8689	0.407	150.7195	0.469	151.6836
0.387	147.7313	1.093	147.8896	0.441	148.8714	0.411	150.7220	0.469	151.6860
0.395	147.7337	1.067	147.8921	0.435	148.8738	0.410	150.7244	0.480	151.6884
0.391	147.7362	1.017	147.8945	0.439	148.8762	0.406	150.7268	0.482	151.6909
0.392	147.7386	0.968	147.8969	0.433	148.8787	0.406	150.7292	0.488	151.6933
0.391	147.7410	0.939	147.8994	0.432	148.8811	0.409	150.7317	0.496	151.6958
0.393	147.7435	0.904	147.9018	0.425	148.8835	0.412	150.7341	0.502	151.6982
0.394	147.7459	0.878	147.9042	0.426	148.8860	0.419	150.7365	0.504	151.7007
0.397	147.7484	0.832	147.9067	0.426	148.8884	0.420	150.7390	0.507	151.7031
0.391	147.7508	0.801	147.9091	0.546	150.5831	0.421	150.7414	0.510	151.7055
0.394	147.7532	0.764	147.9115	0.544	150.5855	0.429	150.7438	0.513	151.7080
0.402	147.7556	0.547	148.7377	0.546	150.5879	0.429	150.7463	0.517	151.7104
0.397	147.7581	0.547	148.7401	0.544	150.5904	0.433	150.7487	0.520	151.7128
0.394	147.7605	0.551	148.7425	0.541	150.5928	0.438	150.7511	0.532	151.7153
0.405	147.7630	0.552	148.7449	0.536	150.5953	0.437	150.7536	0.540	151.7177
0.407	147.7654	0.558	148.7474	0.531	150.5977	0.414	151.5616	0.543	151.7202
0.407	147.7678	0.557	148.7498	0.527	150.6001	0.418	151.5641	0.546	151.7226
0.413	147.7703	0.556	148.7522	0.525	150.6026	0.414	151.5665	0.551	151.7250
0.420	147.7727	0.556	148.7547	0.517	150.6050	0.409	151.5690	0.553	151.7274
0.421	147.7751	0.556	148.7571	0.512	150.6074	0.410	151.5714	0.550	151.7299
0.424	147.7776	0.565	148.7595	0.513	150.6099	0.408	151.5738	0.556	151.7323
0.428	147.7800	0.563	148.7620	0.501	150.6123	0.410	151.5762	0.553	151.7348
0.431	147.7824	0.567	148.7644	0.499	150.6148	0.398	151.5787	0.553	151.7372
0.437	147.7849	0.563	148.7668	0.494	150.6172	0.408	151.5811	0.551	151.7396
0.442	147.7873	0.567	148.7693	0.491	150.6196	0.395	151.5836	0.554	151.7421
0.443	147.7897	0.565	148.7717	0.486	150.6221	0.399	151.5860	0.564	151.7445
0.445	147.7922	0.567	148.7741	0.484	150.6245	0.406	151.5884	0.558	151.7469
0.446	147.7946	0.566	148.7766	0.484	150.6270	0.398	151.5909	0.569	151.7494
0.457	147.7971	0.566	148.7790	0.476	150.6294	0.400	151.5933	0.563	151.7518
0.461	147.7995	0.568	148.7814	0.475	150.6319	0.400	151.5957	0.559	151.7542
0.461	147.8019	0.564	148.7839	0.473	150.6343	0.393	151.5982	0.558	151.7567
0.479	147.8044	0.567	148.7863	0.468	150.6367	0.399	151.6006	0.557	151.7591
0.479	147.8068	0.560	148.7887	0.463	150.6392	0.400	151.6031	0.562	151.7615
0.485	147.8092	0.557	148.7912	0.463	150.6416	0.396	151.6055	0.556	151.7640
0.493	147.8117	0.553	148.7936	0.460	150.6440	0.398	151.6079	0.540	151.7664
0.508	147.8141	0.550	148.7960	0.453	150.6465	0.395	151.6104	0.544	151.7688
0.521	147.8165	0.541	148.7984	0.446	150.6489	0.402	151.6128	0.537	151.7713
0.541	147.8190	0.543	148.8009	0.446	150.6514	0.402	151.6153	0.530	151.7737
0.563	147.8214	0.537	148.8033	0.442	150.6538	0.405	151.6177	0.536	151.7761
0.579	147.8238	0.533	148.8057	0.442	150.6562	0.404	151.6201	0.527	151.7785
0.595	147.8263	0.526	148.8082	0.442	150.6587	0.408	151.6226	0.526	151.7810
0.617	147.8287	0.517	148.8106	0.440	150.6611	0.408	151.6250	0.522	151.7834

Table continued on following pages

Table 2. TYC 1488-693-1 observations, ΔB, ΔV, ΔR$_c$, and ΔI$_c$, variable star minus comparison star.

ΔB	HJD 2457000+	ΔB	HJD 2457000+	ΔB	HJD 2457000+	ΔB	HJD 2457000+	ΔB	HJD 2457000+
0.513	151.7858	0.590	152.5883	1.226	152.6395	0.412	792.9044	0.394	792.9636
0.515	151.7883	0.610	152.5907	1.215	152.6419	0.392	792.9075	0.406	792.9679
0.513	151.7907	0.631	152.5932	1.208	152.6444	0.403	792.9104	0.402	792.9726
0.510	151.7931	0.659	152.5956	0.615	792.8495	0.400	792.9120	0.398	792.9742
0.513	151.7956	0.680	152.5980	0.553	792.8552	0.396	792.9136	0.397	792.9789
0.503	151.7980	0.714	152.6005	0.539	792.8586	0.393	792.9177	0.394	792.9805
0.493	151.8004	0.746	152.6029	0.531	792.8602	0.388	792.9193	0.410	792.9860
0.490	151.8029	0.772	152.6053	0.492	792.8679	0.387	792.9221	0.406	792.9876
0.478	151.8053	0.812	152.6078	0.481	792.8708	0.391	792.9283	0.401	792.9892
0.474	151.8078	0.848	152.6102	0.464	792.8740	0.391	792.9299	0.405	792.9923
0.477	151.8102	0.879	152.6127	0.465	792.8766	0.381	792.9341	0.419	792.9939
0.476	151.8126	0.929	152.6151	0.459	792.8783	0.408	792.9374	0.430	792.9955
0.463	151.8151	0.973	152.6175	0.453	792.8799	0.405	792.9423	0.414	792.9984
0.457	151.8175	1.007	152.6200	0.436	792.8828	0.403	792.9438	0.422	793.0015
0.457	151.8199	1.047	152.6224	0.433	792.8845	0.402	792.9467	0.409	793.0043
0.494	152.5736	1.090	152.6249	0.403	792.8902	0.385	792.9483	0.446	793.0059
0.511	152.5761	1.125	152.6273	0.431	792.8918	0.392	792.9499	0.447	793.0075
0.537	152.5785	1.156	152.6297	0.429	792.8934	0.386	792.9551		
0.540	152.5810	1.188	152.6322	0.410	792.8986	0.389	792.9567		
0.564	152.5834	1.205	152.6346	0.415	792.9002	0.374	792.9604		
0.577	152.5858	1.210	152.6371	0.409	792.9018	0.392	792.9620		

ΔV	HJD 2457000+	ΔV	HJD 2457000+	ΔV	HJD 2457000+	ΔV	HJD 2457000+	ΔV	HJD 2457000+
0.667	113.7154	0.628	113.8119	1.134	113.9082	0.793	117.7823	0.713	117.8696
0.663	113.7178	0.634	113.8143	1.175	113.9106	0.805	117.7845	0.705	117.8718
0.661	113.7202	0.637	113.8167	1.211	113.9130	0.799	117.7866	0.707	117.8740
0.666	113.7226	0.637	113.8191	1.236	113.9154	0.807	117.7888	0.704	117.8761
0.664	113.7250	0.644	113.8215	1.284	113.9179	0.803	117.7910	0.697	117.8783
0.658	113.7274	0.641	113.8239	1.326	113.9203	0.802	117.7932	0.696	117.8805
0.656	113.7298	0.645	113.8263	1.342	113.9227	0.799	117.7954	0.689	117.8827
0.667	113.7323	0.661	113.8287	1.376	113.9251	0.812	117.7975	0.692	117.8849
0.654	113.7347	0.657	113.8311	1.405	113.9275	0.814	117.7997	0.691	117.8870
0.647	113.7371	0.655	113.8335	1.423	113.9299	0.806	117.8019	0.693	117.8892
0.655	113.7395	0.658	113.8359	1.432	113.9323	0.807	117.8041	0.692	117.8914
0.653	113.7419	0.656	113.8383	1.422	113.9347	0.809	117.8063	0.684	117.8936
0.650	113.7444	0.664	113.8408	1.408	113.9371	0.809	117.8085	0.669	117.8958
0.646	113.7468	0.672	113.8432	1.395	113.9395	0.813	117.8107	0.685	117.8979
0.639	113.7492	0.675	113.8456	1.380	113.9419	0.806	117.8129	0.681	117.9001
0.634	113.7516	0.691	113.8480	0.691	117.7277	0.804	117.8151	0.674	117.9023
0.644	113.7540	0.690	113.8504	0.687	117.7299	0.803	117.8173	0.666	117.9045
0.634	113.7564	0.679	113.8528	0.695	117.7321	0.803	117.8194	0.664	117.9067
0.635	113.7589	0.686	113.8552	0.691	117.7343	0.798	117.8216	0.671	117.9089
0.628	113.7613	0.697	113.8576	0.697	117.7365	0.795	117.8238	0.664	117.9110
0.622	113.7637	0.698	113.8600	0.710	117.7386	0.793	117.8260	0.658	117.9132
0.633	113.7661	0.700	113.8624	0.711	117.7408	0.789	117.8282	0.668	117.9154
0.627	113.7685	0.709	113.8649	0.713	117.7430	0.788	117.8303	0.662	117.9176
0.623	113.7709	0.725	113.8673	0.713	117.7452	0.781	117.8325	0.661	117.9198
0.627	113.7733	0.738	113.8697	0.715	117.7474	0.780	117.8347	0.666	117.9219
0.627	113.7757	0.743	113.8721	0.725	117.7495	0.784	117.8369	0.666	117.9241
0.623	113.7782	0.771	113.8745	0.739	117.7517	0.772	117.8391	0.682	117.9241
0.629	113.7806	0.776	113.8769	0.731	117.7539	0.767	117.8412	0.670	117.9263
0.615	113.7830	0.794	113.8793	0.740	117.7561	0.756	117.8434	0.668	117.9285
0.620	113.7854	0.807	113.8817	0.742	117.7583	0.754	117.8456	0.667	117.9307
0.619	113.7878	0.826	113.8841	0.736	117.7605	0.743	117.8478	0.671	117.9328
0.622	113.7902	0.845	113.8865	0.751	117.7626	0.742	117.8500	0.667	117.9350
0.625	113.7926	0.876	113.8889	0.760	117.7648	0.745	117.8521	0.654	117.9372
0.631	113.7950	0.901	113.8914	0.764	117.7670	0.741	117.8543	0.653	117.9394
0.625	113.7974	0.925	113.8938	0.768	117.7692	0.728	117.8565	0.658	117.9416
0.629	113.7998	0.952	113.8962	0.777	117.7714	0.728	117.8587	0.643	136.6598
0.629	113.8022	0.990	113.8986	0.781	117.7736	0.729	117.8609	0.663	136.6622
0.620	113.8046	1.018	113.9010	0.778	117.7757	0.718	117.8630	0.657	136.6647
0.626	113.8071	1.048	113.9034	0.787	117.7779	0.717	117.8652	0.657	136.6671
0.629	113.8095	1.096	113.9058	0.794	117.7801	0.714	117.8674	0.648	136.6696

Table continued on following pages

Samec et al., *JAAVSO Volume 46, 2018*

Table 2. TYC 1488-693-1 observations, ΔB, ΔV, ΔR$_c$, and ΔI$_c$, variable star minus comparison star, cont.

ΔV	HJD 2457000+	ΔV	HJD 2457000+	ΔV	HJD 2457000+	ΔV	HJD 2457000+	ΔV	HJD 2457000+
0.648	136.6720	0.773	136.8309	0.661	147.6785	0.945	147.8369	0.745	148.8187
0.639	136.6745	0.770	136.8333	0.666	147.6810	0.966	147.8393	0.736	148.8212
0.645	136.6769	0.779	136.8357	0.664	147.6834	1.005	147.8417	0.727	148.8236
0.638	136.6793	0.787	136.8382	0.659	147.6859	1.037	147.8442	0.727	148.8260
0.638	136.6818	0.781	136.8406	0.659	147.6883	1.077	147.8466	0.715	148.8285
0.630	136.6842	0.787	136.8430	0.649	147.6907	1.114	147.8491	0.717	148.8309
0.633	136.6867	0.794	136.8455	0.655	147.6931	1.139	147.8515	0.719	148.8333
0.638	136.6891	0.801	136.8479	0.639	147.6956	1.189	147.8539	0.722	148.8358
0.627	136.6916	0.790	136.8504	0.637	147.6980	1.236	147.8564	0.716	148.8382
0.630	136.6940	0.804	136.8528	0.631	147.7005	1.263	147.8588	0.712	148.8406
0.632	136.6965	0.800	136.8553	0.646	147.7029	1.306	147.8612	0.710	148.8430
0.626	136.6990	0.810	136.8577	0.629	147.7054	1.337	147.8637	0.700	148.8455
0.615	136.7014	0.799	136.8601	0.638	147.7078	1.384	147.8661	0.696	148.8479
0.618	136.7039	0.808	136.8626	0.643	147.7102	1.381	147.8685	0.694	148.8503
0.624	136.7063	0.805	136.8650	0.637	147.7127	1.441	147.8710	0.697	148.8528
0.628	136.7088	0.803	136.8674	0.652	147.7151	1.410	147.8734	0.685	148.8552
0.631	136.7112	0.793	136.8699	0.629	147.7175	1.418	147.8759	0.685	148.8576
0.624	136.7137	0.789	136.8723	0.643	147.7200	1.420	147.8783	0.684	148.8601
0.624	136.7161	0.786	136.8748	0.630	147.7224	1.443	147.8807	0.679	148.8625
0.624	136.7186	0.782	136.8772	0.635	147.7248	1.386	147.8832	0.680	148.8649
0.628	136.7210	0.779	136.8797	0.633	147.7273	1.385	147.8856	0.690	148.8674
0.629	136.7234	0.763	136.8821	0.643	147.7297	1.324	147.8880	0.665	148.8698
0.639	136.7259	0.778	136.8846	0.631	147.7321	1.321	147.8905	0.670	148.8722
0.644	136.7283	0.766	136.8870	0.626	147.7346	1.284	147.8929	0.658	148.8746
0.633	136.7308	0.772	136.8894	0.647	147.7370	1.231	147.8953	0.668	148.8771
0.625	136.7332	0.764	136.8919	0.645	147.7395	1.194	147.8978	0.662	148.8795
0.627	136.7357	0.765	136.8943	0.641	147.7419	1.162	147.9002	0.650	148.8819
0.626	136.7381	0.761	136.8967	0.650	147.7443	1.130	147.9026	0.670	148.8844
0.625	136.7406	0.754	136.8992	0.648	147.7468	1.090	147.9051	0.657	148.8868
0.638	136.7430	0.754	136.9016	0.655	147.7492	1.027	147.9075	0.659	148.8892
0.645	136.7455	0.735	136.9041	0.652	147.7516	1.062	147.9099	0.676	148.8917
0.643	136.7479	0.713	136.9065	0.643	147.7541	0.969	147.9124	0.664	148.8941
0.653	136.7504	0.726	136.9090	0.645	147.7565	0.959	147.9148	0.664	148.8965
0.644	136.7528	0.720	136.9114	0.648	147.7589	0.800	148.7409	0.658	148.8990
0.653	136.7552	0.712	136.9138	0.649	147.7614	0.798	148.7434	0.642	148.9014
0.655	136.7577	0.707	136.9163	0.663	147.7638	0.795	148.7458	0.645	148.9038
0.650	136.7601	0.714	136.9187	0.651	147.7662	0.806	148.7482	0.647	148.9063
0.663	136.7626	0.693	136.9211	0.662	147.7687	0.801	148.7506	0.640	148.9087
0.662	136.7650	0.682	136.9236	0.658	147.7711	0.804	148.7531	0.657	148.9111
0.669	136.7675	0.777	147.6151	0.664	147.7736	0.813	148.7555	0.654	148.9136
0.668	136.7699	0.753	147.6176	0.678	147.7760	0.814	148.7579	0.626	148.9160
0.659	136.7723	0.746	147.6200	0.665	147.7784	0.809	148.7604	0.766	150.5839
0.670	136.7748	0.769	147.6225	0.670	147.7809	0.817	148.7628	0.768	150.5863
0.674	136.7772	0.760	147.6249	0.674	147.7833	0.815	148.7652	0.772	150.5888
0.678	136.7796	0.748	147.6273	0.674	147.7857	0.820	148.7677	0.761	150.5912
0.670	136.7821	0.729	147.6298	0.687	147.7882	0.816	148.7701	0.774	150.5937
0.673	136.7845	0.721	147.6322	0.675	147.7906	0.814	148.7726	0.776	150.5961
0.685	136.7870	0.724	147.6346	0.702	147.7930	0.821	148.7750	0.758	150.5986
0.685	136.7894	0.722	147.6371	0.698	147.7955	0.824	148.7774	0.773	150.6010
0.699	136.7918	0.700	147.6395	0.694	147.7979	0.819	148.7799	0.753	150.6034
0.704	136.7943	0.704	147.6420	0.697	147.8003	0.824	148.7823	0.748	150.6059
0.694	136.7967	0.700	147.6444	0.706	147.8028	0.815	148.7847	0.739	150.6083
0.706	136.7991	0.708	147.6468	0.717	147.8052	0.805	148.7872	0.733	150.6107
0.708	136.8016	0.695	147.6493	0.732	147.8076	0.804	148.7896	0.741	150.6132
0.719	136.8040	0.695	147.6517	0.733	147.8101	0.798	148.7920	0.726	150.6156
0.708	136.8065	0.689	147.6541	0.751	147.8125	0.801	148.7944	0.718	150.6181
0.728	136.8089	0.692	147.6566	0.763	147.8150	0.795	148.7969	0.721	150.6205
0.734	136.8113	0.685	147.6590	0.777	147.8174	0.779	148.7993	0.704	150.6229
0.739	136.8138	0.691	147.6615	0.789	147.8198	0.783	148.8017	0.712	150.6254
0.743	136.8162	0.698	147.6639	0.813	147.8223	0.774	148.8042	0.711	150.6278
0.736	136.8186	0.672	147.6664	0.814	147.8247	0.771	148.8066	0.704	150.6303
0.750	136.8211	0.670	147.6688	0.839	147.8271	0.770	148.8090	0.694	150.6327
0.757	136.8235	0.686	147.6712	0.874	147.8296	0.764	148.8114	0.697	150.6351
0.762	136.8260	0.675	147.6737	0.882	147.8320	0.752	148.8139	0.696	150.6376
0.773	136.8284	0.663	147.6761	0.919	147.8344	0.751	148.8163	0.691	150.6400

Table continued on following pages

Table 2. TYC 1488-693-1 observations, ΔB, ΔV, ΔR$_c$, and ΔI$_c$, variable star minus comparison star.

ΔV	HJD 2457000+	ΔV	HJD 2457000+	ΔV	HJD 2457000+	ΔV	HJD 2457000+	ΔV	HJD 2457000+
0.688	150.6425	0.667	151.5674	0.730	151.6869	0.736	151.8062	0.701	792.8834
0.685	150.6449	0.646	151.5698	0.731	151.6893	0.729	151.8086	0.681	792.8939
0.669	150.6473	0.659	151.5722	0.736	151.6918	0.721	151.8110	0.675	792.8991
0.674	150.6498	0.667	151.5747	0.744	151.6942	0.708	151.8135	0.675	792.9048
0.673	150.6522	0.656	151.5771	0.750	151.6966	0.722	151.8159	0.674	792.9064
0.669	150.6546	0.668	151.5795	0.752	151.6991	0.723	151.8183	0.670	792.9081
0.663	150.6571	0.653	151.5820	0.759	151.7015	0.699	151.8208	0.668	792.9109
0.668	150.6595	0.656	151.5844	0.774	151.7039	0.738	152.5745	0.658	792.9125
0.675	150.6620	0.650	151.5868	0.765	151.7064	0.757	152.5769	0.666	792.9141
0.666	150.6644	0.640	151.5893	0.764	151.7088	0.767	152.5794	0.664	792.9166
0.669	150.6668	0.658	151.5917	0.783	151.7112	0.793	152.5818	0.654	792.9182
0.667	150.6693	0.650	151.5942	0.788	151.7137	0.805	152.5843	0.650	792.9225
0.658	150.6717	0.646	151.5966	0.790	151.7161	0.815	152.5867	0.847	792.9229
0.665	150.6741	0.650	151.5990	0.801	151.7186	0.846	152.5891	0.649	792.9241
0.657	150.6766	0.647	151.6015	0.803	151.7210	0.844	152.5916	0.655	792.9288
0.651	150.6790	0.653	151.6039	0.808	151.7234	0.877	152.5940	0.643	792.9304
0.647	150.6815	0.645	151.6064	0.800	151.7259	0.914	152.5965	0.657	792.9320
0.649	150.6839	0.655	151.6088	0.821	151.7283	0.934	152.5989	0.651	792.9347
0.647	150.6863	0.662	151.6112	0.814	151.7307	0.959	152.6013	0.653	792.9363
0.642	150.6888	0.657	151.6137	0.814	151.7332	0.996	152.6038	0.653	792.9427
0.644	150.6912	0.661	151.6161	0.820	151.7356	1.019	152.6062	0.660	792.9443
0.636	150.6936	0.652	151.6185	0.825	151.7380	1.055	152.6086	0.637	792.9472
0.636	150.6961	0.657	151.6210	0.825	151.7405	1.110	152.6111	0.648	792.9488
0.643	150.6985	0.657	151.6234	0.817	151.7429	1.131	152.6135	0.647	792.9504
0.639	150.7010	0.656	151.6259	0.828	151.7453	1.173	152.6160	0.652	792.9540
0.637	150.7034	0.661	151.6283	0.828	151.7478	1.207	152.6184	0.643	792.9556
0.636	150.7058	0.666	151.6308	0.832	151.7502	1.253	152.6208	0.645	792.9572
0.650	150.7082	0.668	151.6332	0.837	151.7526	1.293	152.6233	0.652	792.9609
0.645	150.7107	0.671	151.6356	0.827	151.7551	1.306	152.6257	0.640	792.9625
0.650	150.7131	0.666	151.6381	0.833	151.7575	1.373	152.6282	0.644	792.9700
0.645	150.7155	0.672	151.6405	0.829	151.7599	1.388	152.6306	0.655	792.9731
0.647	150.7180	0.671	151.6430	0.815	151.7624	1.424	152.6330	0.646	792.9747
0.645	150.7204	0.685	151.6454	0.820	151.7648	1.435	152.6355	0.657	792.9794
0.651	150.7228	0.678	151.6478	0.813	151.7672	1.432	152.6379	0.665	792.9810
0.645	150.7252	0.687	151.6503	0.806	151.7697	1.435	152.6404	0.665	792.9826
0.643	150.7277	0.678	151.6527	0.795	151.7721	1.435	152.6428	0.665	792.9881
0.641	150.7301	0.691	151.6552	0.777	151.7745	1.402	152.6452	0.846	792.9884
0.658	150.7325	0.697	151.6576	0.792	151.7770	0.874	792.8500	0.669	792.9897
0.653	150.7350	0.692	151.6600	0.785	151.7794	0.861	792.8516	0.673	792.9928
0.657	150.7374	0.694	151.6625	0.782	151.7818	0.834	792.8557	0.672	792.9960
0.675	150.7398	0.697	151.6649	0.771	151.7842	0.796	792.8591	0.672	792.9988
0.664	150.7422	0.697	151.6674	0.772	151.7867	0.783	792.8607	0.677	793.0004
0.668	150.7447	0.697	151.6698	0.773	151.7891	0.768	792.8623	0.682	793.0020
0.673	150.7471	0.705	151.6722	0.765	151.7915	0.766	792.8652	0.658	793.0064
0.682	150.7495	0.704	151.6747	0.752	151.7940	0.766	792.8668	0.618	793.0080
0.696	150.7520	0.715	151.6771	0.744	151.7964	0.752	792.8683		
0.680	150.7544	0.717	151.6796	0.754	151.7989	0.735	792.8729		
0.647	151.5625	0.718	151.6820	0.732	151.8013	0.719	792.8745		
0.662	151.5649	0.721	151.6844	0.728	151.8037	0.722	792.8772		

Table continued on following pages

Table 2. TYC 1488-693-1 observations, ΔB, ΔV, ΔR$_c$, and ΔI$_c$, variable star minus comparison star, cont.

ΔR$_c$	HJD 2457000+	ΔR$_c$	HJD 2457000+	ΔR$_c$	HJD 2457000+	ΔR$_c$	HJD 2457000+	ΔR$_c$	HJD 2457000+
0.908	113.7157	0.981	113.8724	1.056	117.8066	0.911	136.6651	1.002	136.8239
0.907	113.7181	0.991	113.8748	1.055	117.8088	0.899	136.6675	1.012	136.8263
0.896	113.7206	1.011	113.8772	1.054	117.8110	0.891	136.6699	1.010	136.8288
0.904	113.7230	1.021	113.8797	1.057	117.8132	0.889	136.6724	1.024	136.8312
0.901	113.7254	1.043	113.8821	1.062	117.8154	0.893	136.6748	1.030	136.8336
0.899	113.7278	1.062	113.8845	1.054	117.8175	0.886	136.6772	1.027	136.8361
0.898	113.7302	1.086	113.8869	1.048	117.8197	0.881	136.6797	1.036	136.8385
0.898	113.7326	1.105	113.8893	1.042	117.8219	0.888	136.6821	1.036	136.8410
0.885	113.7350	1.129	113.8917	1.037	117.8241	0.878	136.6846	1.041	136.8434
0.890	113.7375	1.158	113.8941	1.035	117.8263	0.885	136.6870	1.050	136.8459
0.885	113.7399	1.187	113.8965	1.037	117.8285	0.875	136.6895	1.054	136.8483
0.882	113.7423	1.225	113.8989	1.033	117.8306	0.878	136.6920	1.055	136.8507
0.886	113.7447	1.247	113.9013	1.023	117.8328	0.873	136.6944	1.054	136.8532
0.881	113.7471	1.282	113.9037	1.017	117.8350	0.867	136.6969	1.062	136.8556
0.883	113.7495	1.325	113.9062	1.009	117.8372	0.861	136.6993	1.062	136.8580
0.883	113.7520	1.354	113.9086	1.007	117.8394	0.866	136.7018	1.056	136.8605
0.876	113.7544	1.396	113.9110	1.007	117.8415	0.865	136.7042	1.053	136.8629
0.870	113.7568	1.427	113.9134	0.998	117.8437	0.864	136.7067	1.049	136.8654
0.872	113.7592	1.471	113.9158	0.986	117.8459	0.873	136.7091	1.062	136.8678
0.868	113.7616	1.496	113.9182	0.979	117.8481	0.871	136.7116	1.044	136.8703
0.862	113.7640	1.530	113.9206	0.973	117.8503	0.876	136.7140	1.045	136.8727
0.863	113.7664	1.562	113.9230	0.968	117.8524	0.873	136.7165	1.036	136.8751
0.862	113.7689	1.590	113.9254	0.962	117.8546	0.865	136.7189	1.032	136.8776
0.861	113.7713	1.614	113.9279	0.957	117.8568	0.861	136.7214	1.019	136.8800
0.865	113.7737	1.620	113.9302	0.955	117.8590	0.865	136.7238	1.023	136.8825
0.858	113.7761	1.630	113.9327	0.956	117.8612	0.868	136.7263	1.022	136.8849
0.866	113.7785	1.622	113.9351	0.950	117.8633	0.876	136.7287	1.017	136.8873
0.864	113.7809	1.613	113.9375	0.942	117.8655	0.876	136.7311	1.020	136.8898
0.865	113.7833	1.598	113.9399	0.945	117.8677	0.876	136.7336	1.013	136.8922
0.862	113.7857	0.914	117.7280	0.934	117.8699	0.878	136.7360	1.006	136.8947
0.870	113.7881	0.924	117.7302	0.932	117.8721	0.876	136.7385	0.989	136.8971
0.861	113.7905	0.916	117.7324	0.926	117.8743	0.872	136.7409	0.998	136.8995
0.870	113.7929	0.925	117.7346	0.922	117.8764	0.881	136.7434	0.983	136.9020
0.871	113.7954	0.929	117.7368	0.912	117.8786	0.877	136.7458	0.974	136.9044
0.866	113.7978	0.935	117.7389	0.919	117.8808	0.882	136.7483	0.965	136.9069
0.866	113.8002	0.932	117.7411	0.918	117.8830	0.879	136.7507	0.961	136.9093
0.869	113.8026	0.936	117.7433	0.919	117.8852	0.892	136.7532	0.964	136.9118
0.863	113.8050	0.946	117.7455	0.914	117.8873	0.883	136.7556	0.938	136.9142
0.867	113.8074	0.955	117.7477	0.907	117.8895	0.895	136.7581	0.953	136.9166
0.879	113.8098	0.955	117.7498	0.909	117.8917	0.884	136.7605	0.938	136.9191
0.876	113.8122	0.958	117.7520	0.913	117.8939	0.900	136.7629	0.931	136.9215
0.880	113.8146	0.957	117.7542	0.908	117.8961	0.899	136.7654	0.948	136.9239
0.880	113.8170	0.966	117.7564	0.906	117.8982	0.900	136.7678	0.914	136.9264
0.882	113.8194	0.971	117.7586	0.908	117.9004	0.897	136.7702	0.990	147.6155
0.890	113.8218	0.972	117.7607	0.906	117.9026	0.901	136.7727	0.994	147.6179
0.881	113.8242	0.987	117.7629	0.898	117.9048	0.900	136.7751	0.987	147.6204
0.894	113.8267	0.983	117.7651	0.900	117.9070	0.908	136.7776	0.980	147.6228
0.894	113.8291	0.998	117.7673	0.901	117.9091	0.904	136.7800	0.977	147.6253
0.891	113.8315	0.998	117.7695	0.890	117.9113	0.909	136.7824	0.965	147.6277
0.897	113.8339	1.013	117.7717	0.886	117.9135	0.913	136.7849	0.949	147.6301
0.909	113.8363	1.022	117.7738	0.889	117.9157	0.914	136.7873	0.948	147.6326
0.904	113.8387	1.023	117.7760	0.892	117.9179	0.927	136.7897	0.943	147.6350
0.904	113.8411	1.034	117.7782	0.890	117.9200	0.924	136.7922	0.942	147.6374
0.913	113.8435	1.038	117.7804	0.890	117.9222	0.922	136.7946	0.940	147.6399
0.917	113.8459	1.034	117.7826	0.888	117.9244	0.939	136.7971	0.932	147.6423
0.921	113.8483	1.042	117.7847	0.898	117.9266	0.943	136.7995	0.930	147.6447
0.919	113.8507	1.052	117.7869	0.908	117.9288	0.944	136.8019	0.917	147.6472
0.923	113.8531	1.043	117.7891	0.899	117.9310	0.949	136.8044	0.928	147.6496
0.931	113.8556	1.050	117.7913	0.895	117.9331	0.968	136.8068	0.920	147.6521
0.929	113.8580	1.047	117.7935	0.897	117.9353	0.961	136.8093	0.920	147.6545
0.932	113.8604	1.048	117.7957	0.901	117.9375	0.967	136.8117	0.914	147.6569
0.948	113.8628	1.057	117.7978	0.894	117.9397	0.975	136.8141	0.907	147.6594
0.952	113.8652	1.058	117.8000	0.894	117.9419	0.976	136.8166	0.911	147.6618
0.962	113.8676	1.057	117.8022	0.904	136.6602	0.988	136.8190	0.916	147.6643
0.983	113.8700	1.063	117.8044	0.897	136.6626	0.990	136.8215	0.900	147.6667

Table continued on following pages

Table 2. TYC 1488-693-1 observations, ΔB, ΔV, ΔR$_c$, and ΔI$_c$, variable star minus comparison star, cont.

ΔR$_c$	HJD 2457000+	ΔR$_c$	HJD 2457000+	ΔR$_c$	HJD 2457000+	ΔR$_c$	HJD 2457000+	ΔR$_c$	HJD 2457000+
0.891	147.6692	1.072	147.8275	0.980	148.8118	0.924	150.6355	0.883	151.5994
0.897	147.6716	1.107	147.8299	0.990	148.8142	0.923	150.6379	0.881	151.6018
0.890	147.6740	1.119	147.8323	0.968	148.8167	0.913	150.6404	0.883	151.6043
0.889	147.6765	1.138	147.8348	0.962	148.8191	0.911	150.6428	0.888	151.6067
0.892	147.6789	1.178	147.8372	0.963	148.8215	0.907	150.6452	0.880	151.6091
0.890	147.6813	1.203	147.8397	0.945	148.8239	0.904	150.6477	0.885	151.6116
0.888	147.6838	1.226	147.8421	0.940	148.8264	0.900	150.6501	0.891	151.6140
0.884	147.6862	1.256	147.8445	0.949	148.8288	0.898	150.6526	0.881	151.6164
0.888	147.6886	1.292	147.8470	0.946	148.8312	0.902	150.6550	0.893	151.6189
0.880	147.6911	1.334	147.8494	0.946	148.8337	0.900	150.6574	0.887	151.6213
0.885	147.6935	1.387	147.8519	0.935	148.8361	0.899	150.6599	0.891	151.6238
0.876	147.6959	1.418	147.8543	0.939	148.8385	0.897	150.6623	0.888	151.6262
0.871	147.6984	1.448	147.8567	0.930	148.8410	0.896	150.6647	0.892	151.6287
0.874	147.7008	1.486	147.8592	0.927	148.8434	0.884	150.6672	0.886	151.6311
0.876	147.7033	1.515	147.8616	0.928	148.8458	0.897	150.6696	0.897	151.6335
0.875	147.7057	1.549	147.8640	0.922	148.8482	0.894	150.6721	0.899	151.6360
0.865	147.7081	1.577	147.8665	0.915	148.8507	0.885	150.6745	0.893	151.6384
0.870	147.7106	1.611	147.8689	0.902	148.8531	0.886	150.6769	0.893	151.6409
0.863	147.7130	1.609	147.8713	0.914	148.8555	0.884	150.6794	0.900	151.6433
0.865	147.7154	1.614	147.8738	0.907	148.8580	0.884	150.6818	0.896	151.6458
0.866	147.7179	1.613	147.8762	0.901	148.8604	0.882	150.6842	0.908	151.6482
0.871	147.7203	1.620	147.8786	0.898	148.8628	0.870	150.6867	0.907	151.6506
0.868	147.7227	1.626	147.8811	0.896	148.8653	0.872	150.6891	0.916	151.6531
0.870	147.7252	1.591	147.8835	0.899	148.8677	0.873	150.6916	0.914	151.6555
0.866	147.7276	1.584	147.8860	0.901	148.8701	0.877	150.6940	0.917	151.6580
0.874	147.7301	1.542	147.8884	0.892	148.8726	0.872	150.6964	0.915	151.6604
0.870	147.7325	1.516	147.8908	0.898	148.8750	0.872	150.6989	0.921	151.6628
0.861	147.7349	1.479	147.8933	0.896	148.8774	0.878	150.7013	0.918	151.6653
0.867	147.7374	1.457	147.8957	0.884	148.8799	0.872	150.7037	0.917	151.6677
0.873	147.7398	1.397	147.8981	0.883	148.8823	0.872	150.7062	0.919	151.6701
0.868	147.7422	1.369	147.9005	0.889	148.8847	0.875	150.7086	0.921	151.6726
0.866	147.7447	1.335	147.9030	0.876	148.8872	0.879	150.7110	0.923	151.6750
0.871	147.7471	1.296	147.9054	0.883	148.8896	0.874	150.7134	0.941	151.6775
0.876	147.7495	1.238	147.9079	0.876	148.8920	0.887	150.7159	0.938	151.6799
0.880	147.7520	1.229	147.9103	0.878	148.8945	0.883	150.7183	0.938	151.6823
0.880	147.7544	1.176	147.9127	0.886	148.8969	0.879	150.7207	0.943	151.6848
0.888	147.7569	1.151	147.9151	0.900	148.8993	0.872	150.7232	0.954	151.6872
0.885	147.7593	1.032	148.7437	0.893	148.9018	0.883	150.7256	0.955	151.6897
0.890	147.7617	1.034	148.7461	0.871	148.9042	0.881	150.7280	0.973	151.6921
0.890	147.7642	1.035	148.7486	0.865	148.9066	0.886	150.7305	0.966	151.6945
0.889	147.7666	1.045	148.7510	0.881	148.9091	0.891	150.7329	0.977	151.6970
0.896	147.7690	1.047	148.7534	0.865	148.9115	0.887	150.7353	0.983	151.6994
0.895	147.7715	1.055	148.7559	0.866	148.9139	0.891	150.7378	0.985	151.7018
0.889	147.7739	1.055	148.7583	0.876	148.9164	0.894	150.7402	0.996	151.7043
0.911	147.7764	1.059	148.7607	1.028	150.5843	0.898	150.7426	0.994	151.7067
0.902	147.7788	1.051	148.7632	1.007	150.5867	0.918	150.7450	1.001	151.7092
0.911	147.7812	1.054	148.7656	1.028	150.5891	0.903	150.7475	1.017	151.7116
0.921	147.7837	1.058	148.7680	1.005	150.5916	0.918	150.7499	1.018	151.7140
0.918	147.7861	1.055	148.7705	1.004	150.5940	0.917	150.7523	1.023	151.7165
0.912	147.7885	1.058	148.7729	1.000	150.5965	0.881	150.7548	1.043	151.7189
0.929	147.7910	1.063	148.7753	0.984	150.5989	0.879	151.5628	1.041	151.7214
0.922	147.7934	1.060	148.7778	0.982	150.6013	0.897	151.5653	1.047	151.7238
0.928	147.7958	1.063	148.7802	0.986	150.6038	0.890	151.5677	1.062	151.7262
0.934	147.7983	1.056	148.7826	0.980	150.6062	0.906	151.5702	1.055	151.7287
0.936	147.8007	1.056	148.7851	0.974	150.6086	0.897	151.5726	1.063	151.7311
0.936	147.8031	1.045	148.7875	0.960	150.6111	0.910	151.5750	1.064	151.7335
0.938	147.8056	1.034	148.7899	0.958	150.6135	0.895	151.5775	1.073	151.7360
0.955	147.8080	1.034	148.7924	0.950	150.6160	0.896	151.5799	1.074	151.7384
0.968	147.8104	1.041	148.7948	0.941	150.6184	0.879	151.5823	1.059	151.7408
0.976	147.8129	1.029	148.7972	0.938	150.6209	0.892	151.5848	1.071	151.7433
0.981	147.8153	1.023	148.7997	0.940	150.6233	0.900	151.5872	1.068	151.7457
1.000	147.8177	1.014	148.8021	0.935	150.6257	0.878	151.5896	1.067	151.7481
1.014	147.8202	1.010	148.8045	0.931	150.6282	0.896	151.5921	1.071	151.7506
1.036	147.8226	1.002	148.8069	0.923	150.6306	0.880	151.5945	1.076	151.7530
1.048	147.8250	0.989	148.8094	0.926	150.6330	0.877	151.5969	1.073	151.7554

Table continued on following pages

Table 2. TYC 1488-693-1 observations, ΔB, ΔV, ΔR$_c$, and ΔI$_c$, variable star minus comparison star, cont.

ΔR$_c$	HJD 2457000+	ΔR$_c$	HJD 2457000+	ΔR$_c$	HJD 2457000+	ΔR$_c$	HJD 2457000+	ΔR$_c$	HJD 2457000+
1.073	151.7579	0.927	151.8163	1.502	152.6236	0.902	792.8994	0.868	792.9628
1.060	151.7603	0.934	151.8187	1.545	152.6261	0.884	792.9052	0.871	792.9655
1.053	151.7627	0.921	151.8212	1.568	152.6285	0.882	792.9067	0.864	792.9687
1.057	151.7652	0.909	151.8236	1.606	152.6309	0.881	792.9095	0.868	792.9718
1.042	151.7676	0.980	152.5748	1.627	152.6334	0.896	792.9112	0.867	792.9734
1.041	151.7700	0.994	152.5773	1.638	152.6358	0.875	792.9128	0.860	792.9750
1.031	151.7724	1.001	152.5797	1.660	152.6383	0.880	792.9153	0.864	792.9781
1.035	151.7749	1.013	152.5822	1.640	152.6407	0.881	792.9169	0.872	792.9797
1.026	151.7773	1.032	152.5846	1.637	152.6431	0.883	792.9185	0.874	792.9813
1.020	151.7798	1.048	152.5871	1.580	152.6456	0.875	792.9212	0.871	792.9852
1.017	151.7822	1.057	152.5895	1.101	792.8487	0.872	792.9229	0.880	792.9868
1.010	151.7846	1.093	152.5919	1.090	792.8503	0.871	792.9244	0.871	792.9884
0.994	151.7871	1.110	152.5944	1.064	792.8519	0.871	792.9307	0.879	792.9915
0.986	151.7895	1.128	152.5968	0.984	792.8638	0.862	792.9333	0.877	792.9931
0.988	151.7919	1.159	152.5992	0.951	792.8716	0.874	792.9349	0.883	792.9947
0.994	151.7943	1.186	152.6017	0.940	792.8732	0.874	792.9366	0.885	792.9975
0.975	151.7968	1.220	152.6041	0.938	792.8759	0.873	792.9399	0.884	792.9991
0.964	151.7992	1.250	152.6066	0.941	792.8774	0.869	792.9475	0.886	793.0007
0.972	151.8017	1.289	152.6090	0.927	792.8790	0.867	792.9491	0.871	793.0035
0.963	151.8041	1.319	152.6114	0.919	792.8836	0.864	792.9527	0.879	793.0051
0.956	151.8065	1.353	152.6139	0.923	792.8853	0.862	792.9543	0.883	793.0067
0.948	151.8090	1.396	152.6163	0.909	792.8926	0.858	792.9559		
0.940	151.8114	1.434	152.6187	0.905	792.8951	0.868	792.9596		
0.930	151.8138	1.470	152.6212	0.891	792.8978	0.849	792.9612		

ΔI$_c$	HJD 2457000+	ΔI$_c$	HJD 2457000+	ΔI$_c$	HJD 2457000+	ΔI$_c$	HJD 2457000+	ΔI$_c$	HJD 2457000+
1.084	113.7257	1.055	113.8126	1.434	113.9017	1.202	117.7698	1.153	117.8505
1.077	113.7281	1.066	113.8150	1.464	113.9041	1.210	117.7720	1.147	117.8527
1.081	113.7306	1.065	113.8174	1.500	113.9065	1.217	117.7741	1.134	117.8549
1.076	113.7330	1.073	113.8198	1.528	113.9089	1.230	117.7763	1.133	117.8571
1.072	113.7354	1.076	113.8222	1.562	113.9113	1.232	117.7785	1.134	117.8593
1.062	113.7378	1.070	113.8246	1.588	113.9137	1.242	117.7807	1.121	117.8615
1.065	113.7402	1.068	113.8270	1.643	113.9161	1.241	117.7829	1.128	117.8636
1.056	113.7426	1.083	113.8294	1.666	113.9186	1.250	117.7850	1.114	117.8658
1.065	113.7451	1.092	113.8318	1.696	113.9210	1.250	117.7872	1.120	117.8680
1.065	113.7475	1.087	113.8342	1.714	113.9234	1.259	117.7894	1.104	117.8702
1.061	113.7499	1.093	113.8366	1.741	113.9258	1.262	117.7916	1.113	117.8724
1.062	113.7523	1.097	113.8391	1.766	113.9282	1.267	117.7938	1.091	117.8745
1.058	113.7547	1.099	113.8439	1.774	113.9306	1.261	117.7959	1.096	117.8767
1.053	113.7572	1.111	113.8463	1.769	113.9330	1.271	117.7981	1.100	117.8789
1.053	113.7596	1.104	113.8487	1.755	113.9354	1.269	117.8003	1.096	117.8811
1.048	113.7620	1.101	113.8511	1.748	113.9378	1.272	117.8025	1.093	117.8833
1.053	113.7644	1.110	113.8535	1.748	113.9402	1.265	117.8047	1.079	117.8854
1.054	113.7668	1.119	113.8559	1.098	117.7283	1.264	117.8069	1.092	117.8876
1.054	113.7692	1.118	113.8583	1.097	117.7305	1.270	117.8091	1.087	117.8898
1.049	113.7716	1.117	113.8607	1.108	117.7327	1.255	117.8113	1.089	117.8920
1.048	113.7740	1.136	113.8631	1.100	117.7349	1.269	117.8135	1.093	117.8942
1.046	113.7764	1.135	113.8656	1.111	117.7370	1.268	117.8157	1.084	117.8985
1.052	113.7788	1.147	113.8680	1.103	117.7392	1.255	117.8178	1.092	117.9007
1.052	113.7812	1.160	113.8704	1.112	117.7414	1.248	117.8200	1.083	117.9029
1.045	113.7837	1.170	113.8728	1.123	117.7436	1.245	117.8222	1.080	117.9051
1.048	113.7861	1.178	113.8752	1.126	117.7458	1.236	117.8244	1.069	117.9073
1.047	113.7885	1.195	113.8776	1.138	117.7479	1.234	117.8266	1.075	117.9094
1.049	113.7909	1.209	113.8800	1.134	117.7501	1.242	117.8288	1.057	117.9116
1.045	113.7933	1.231	113.8824	1.149	117.7523	1.223	117.8309	1.071	117.9138
1.052	113.7957	1.248	113.8848	1.153	117.7545	1.215	117.8331	1.073	117.9160
1.055	113.7981	1.268	113.8872	1.158	117.7567	1.206	117.8353	1.070	117.9182
1.052	113.8005	1.295	113.8896	1.167	117.7589	1.209	117.8375	1.078	117.9203
1.061	113.8029	1.312	113.8921	1.166	117.7610	1.202	117.8397	1.071	117.9225
1.060	113.8053	1.345	113.8945	1.179	117.7632	1.197	117.8418	1.070	117.9247
1.057	113.8077	1.367	113.8969	1.182	117.7654	1.179	117.8440	1.066	117.9269
1.056	113.8102	1.401	113.8993	1.189	117.7676	1.169	117.8484	1.069	117.9291

Table continued on following pages

Table 2. TYC 1488-693-1 observations, ΔB, ΔV, ΔR$_c$, and ΔI$_c$, variable star minus comparison star, cont.

ΔI$_c$	HJD 2457000+	ΔI$_c$	HJD 2457000+	ΔI$_c$	HJD 2457000+	ΔI$_c$	HJD 2457000+	ΔI$_c$	HJD 2457000+
1.074	117.9312	1.141	136.8121	1.099	147.6573	1.204	147.8205	1.217	148.7976
1.066	117.9334	1.160	136.8145	1.091	147.6597	1.220	147.8230	1.219	148.8000
1.074	117.9356	1.164	136.8169	1.099	147.6622	1.248	147.8254	1.213	148.8024
1.082	117.9378	1.173	136.8194	1.092	147.6646	1.263	147.8278	1.206	148.8049
1.077	117.9400	1.182	136.8218	1.085	147.6671	1.288	147.8303	1.188	148.8073
1.081	117.9421	1.190	136.8243	1.079	147.6695	1.296	147.8327	1.183	148.8097
1.085	136.6605	1.195	136.8267	1.075	147.6720	1.332	147.8351	1.175	148.8122
1.081	136.6630	1.209	136.8291	1.077	147.6744	1.352	147.8376	1.168	148.8146
1.090	136.6654	1.205	136.8316	1.076	147.6768	1.383	147.8400	1.160	148.8170
1.086	136.6678	1.221	136.8340	1.070	147.6793	1.416	147.8424	1.150	148.8194
1.073	136.6703	1.230	136.8364	1.072	147.6817	1.453	147.8449	1.137	148.8219
1.077	136.6727	1.229	136.8389	1.075	147.6841	1.475	147.8473	1.136	148.8243
1.063	136.6752	1.245	136.8413	1.074	147.6866	1.500	147.8498	1.134	148.8267
1.067	136.6776	1.250	136.8438	1.066	147.6890	1.546	147.8522	1.125	148.8292
1.076	136.6801	1.258	136.8462	1.063	147.6914	1.592	147.8547	1.123	148.8316
1.068	136.6825	1.261	136.8486	1.067	147.6939	1.609	147.8571	1.116	148.8340
1.070	136.6849	1.254	136.8511	1.062	147.6988	1.653	147.8595	1.126	148.8365
1.059	136.6874	1.273	136.8535	1.068	147.7012	1.687	147.8619	1.114	148.8389
1.060	136.6899	1.253	136.8560	1.056	147.7036	1.714	147.8644	1.115	148.8413
1.064	136.6923	1.258	136.8584	1.061	147.7061	1.732	147.8668	1.114	148.8438
1.054	136.6948	1.269	136.8608	1.057	147.7085	1.754	147.8693	1.104	148.8462
1.042	136.6972	1.256	136.8633	1.051	147.7109	1.772	147.8717	1.098	148.8486
1.051	136.6997	1.260	136.8657	1.048	147.7134	1.769	147.8741	1.092	148.8510
1.056	136.7021	1.260	136.8682	1.049	147.7158	1.765	147.8766	1.092	148.8535
1.053	136.7095	1.258	136.8706	1.051	147.7182	1.788	147.8790	1.091	148.8559
1.047	136.7119	1.249	136.8731	1.065	147.7231	1.781	147.8814	1.088	148.8583
1.053	136.7168	1.229	136.8755	1.051	147.7255	1.737	147.8839	1.094	148.8608
1.054	136.7193	1.231	136.8779	1.047	147.7280	1.722	147.8863	1.089	148.8632
1.051	136.7217	1.222	136.8804	1.050	147.7304	1.728	147.8887	1.081	148.8656
1.061	136.7242	1.229	136.8828	1.048	147.7329	1.669	147.8912	1.089	148.8681
1.063	136.7266	1.226	136.8853	1.060	147.7353	1.631	147.8936	1.087	148.8705
1.060	136.7291	1.219	136.8877	1.049	147.7377	1.586	147.8960	1.074	148.8729
1.067	136.7315	1.211	136.8901	1.054	147.7402	1.563	147.8985	1.090	148.8754
1.067	136.7340	1.202	136.8926	1.060	147.7426	1.542	147.9009	1.075	148.8778
1.052	136.7364	1.188	136.8950	1.060	147.7450	1.493	147.9033	1.074	148.8802
1.057	136.7388	1.189	136.8975	1.049	147.7475	1.460	147.9058	1.074	148.8827
1.059	136.7413	1.181	136.8999	1.067	147.7499	1.432	147.9082	1.086	148.8851
1.055	136.7437	1.155	136.9023	1.072	147.7523	1.410	147.9106	1.056	148.8875
1.062	136.7462	1.171	136.9048	1.067	147.7548	1.375	147.9131	1.066	148.8900
1.057	136.7486	1.140	136.9072	1.072	147.7572	1.346	147.9155	1.070	148.8924
1.060	136.7511	1.140	136.9097	1.074	147.7596	1.299	147.9180	1.060	148.8948
1.054	136.7535	1.150	136.9121	1.092	147.7621	1.218	148.7392	1.078	148.8972
1.067	136.7560	1.121	136.9145	1.081	147.7645	1.229	148.7416	1.069	148.8997
1.060	136.7584	1.112	136.9170	1.082	147.7670	1.231	148.7441	1.060	148.9021
1.064	136.7609	1.066	136.9194	1.075	147.7694	1.240	148.7465	1.068	148.9045
1.083	136.7633	1.065	136.9219	1.093	147.7718	1.251	148.7489	1.068	148.9070
1.059	136.7657	1.124	136.9243	1.083	147.7743	1.247	148.7514	1.057	148.9094
1.067	136.7682	1.068	136.9267	1.091	147.7767	1.252	148.7538	1.081	148.9118
1.074	136.7706	1.194	147.6158	1.088	147.7791	1.254	148.7562	1.056	148.9143
1.079	136.7730	1.175	147.6183	1.102	147.7816	1.259	148.7587	1.039	148.9167
1.082	136.7755	1.171	147.6207	1.103	147.7840	1.281	148.7611	1.207	150.5846
1.066	136.7779	1.178	147.6232	1.096	147.7864	1.271	148.7635	1.209	150.5871
1.077	136.7804	1.147	147.6256	1.110	147.7889	1.271	148.7660	1.185	150.5895
1.094	136.7828	1.139	147.6280	1.118	147.7913	1.273	148.7684	1.203	150.5919
1.091	136.7852	1.150	147.6305	1.116	147.7937	1.272	148.7708	1.197	150.5944
1.092	136.7877	1.138	147.6329	1.117	147.7962	1.261	148.7733	1.174	150.5968
1.093	136.7901	1.133	147.6353	1.122	147.7986	1.275	148.7757	1.179	150.5993
1.091	136.7925	1.123	147.6378	1.115	147.8011	1.266	148.7781	1.191	150.6017
1.096	136.7950	1.119	147.6402	1.132	147.8035	1.258	148.7806	1.163	150.6041
1.108	136.7974	1.113	147.6427	1.124	147.8059	1.270	148.7830	1.164	150.6066
1.123	136.7999	1.111	147.6451	1.135	147.8084	1.259	148.7854	1.143	150.6090
1.115	136.8023	1.104	147.6475	1.151	147.8108	1.257	148.7879	1.141	150.6114
1.121	136.8047	1.091	147.6500	1.162	147.8132	1.244	148.7903	1.137	150.6139
1.126	136.8072	1.092	147.6524	1.175	147.8157	1.241	148.7927	1.120	150.6163
1.140	136.8096	1.096	147.6549	1.188	147.8181	1.236	148.7952	1.126	150.6188

Table continued on next page

Table 2. TYC 1488-693-1 observations, ΔB, ΔV, ΔR$_c$, and ΔI$_c$, variable star minus comparison star, cont.

ΔI$_c$	HJD 2457000+	ΔI$_c$	HJD 2457000+	ΔI$_c$	HJD 2457000+	ΔI$_c$	HJD 2457000+	ΔI$_c$	HJD 2457000+
1.116	150.6212	1.076	150.7381	1.095	151.6656	1.205	151.7825	1.143	792.8702
1.113	150.6236	1.087	150.7405	1.098	151.6681	1.203	151.7850	1.130	792.8718
1.101	150.6261	1.089	150.7430	1.090	151.6705	1.190	151.7874	1.141	792.8735
1.107	150.6285	1.086	150.7454	1.100	151.6729	1.189	151.7898	1.127	792.8762
1.104	150.6310	1.086	150.7478	1.104	151.6754	1.177	151.7923	1.114	792.8777
1.101	150.6334	1.113	150.7503	1.109	151.6778	1.169	151.7947	1.101	792.8823
1.105	150.6358	1.101	150.7527	1.115	151.6803	1.168	151.7971	1.115	792.8839
1.098	150.6383	1.081	151.5632	1.115	151.6827	1.162	151.7996	1.120	792.8855
1.098	150.6407	1.088	151.5656	1.122	151.6851	1.151	151.8020	1.064	792.8896
1.092	150.6432	1.089	151.5681	1.133	151.6876	1.143	151.8045	1.075	792.8912
1.095	150.6456	1.090	151.5705	1.137	151.6900	1.150	151.8069	1.092	792.8929
1.082	150.6480	1.087	151.5729	1.141	151.6925	1.129	151.8093	1.091	792.8954
1.082	150.6505	1.076	151.5754	1.151	151.6949	1.127	151.8118	1.079	792.9039
1.076	150.6529	1.066	151.5778	1.164	151.6973	1.131	151.8142	1.068	792.9070
1.079	150.6553	1.068	151.5827	1.163	151.6998	1.110	151.8166	1.064	792.9099
1.082	150.6578	1.081	151.5851	1.169	151.7022	1.160	152.5752	1.063	792.9131
1.077	150.6602	1.078	151.5876	1.177	151.7047	1.180	152.5777	1.077	792.9156
1.076	150.6627	1.081	151.5900	1.186	151.7071	1.187	152.5801	1.063	792.9188
1.080	150.6651	1.073	151.5924	1.201	151.7095	1.199	152.5825	1.060	792.9215
1.074	150.6675	1.070	151.5949	1.206	151.7120	1.230	152.5850	1.062	792.9232
1.064	150.6700	1.070	151.5973	1.216	151.7144	1.242	152.5874	1.065	792.9278
1.065	150.6724	1.067	151.5997	1.231	151.7168	1.252	152.5898	1.050	792.9310
1.063	150.6749	1.074	151.6022	1.238	151.7193	1.281	152.5923	1.056	792.9337
1.065	150.6773	1.068	151.6046	1.241	151.7217	1.306	152.5947	1.048	792.9401
1.068	150.6797	1.074	151.6071	1.258	151.7241	1.317	152.5972	1.060	792.9417
1.070	150.6822	1.077	151.6095	1.263	151.7266	1.347	152.5996	1.050	792.9434
1.061	150.6846	1.066	151.6119	1.261	151.7290	1.380	152.6020	1.053	792.9462
1.062	150.6870	1.068	151.6144	1.279	151.7314	1.402	152.6045	1.054	792.9494
1.059	150.6895	1.066	151.6168	1.271	151.7339	1.433	152.6069	1.040	792.9546
1.055	150.6919	1.069	151.6192	1.270	151.7363	1.453	152.6093	1.048	792.9562
1.054	150.6944	1.065	151.6217	1.274	151.7387	1.511	152.6118	1.048	792.9599
1.056	150.6968	1.063	151.6241	1.275	151.7412	1.530	152.6142	1.042	792.9615
1.062	150.6992	1.079	151.6266	1.292	151.7436	1.575	152.6167	1.043	792.9658
1.061	150.7017	1.076	151.6290	1.277	151.7461	1.598	152.6191	1.053	792.9673
1.064	150.7041	1.081	151.6315	1.283	151.7485	1.644	152.6215	1.049	792.9720
1.068	150.7065	1.075	151.6339	1.283	151.7509	1.675	152.6240	1.042	792.9753
1.061	150.7090	1.079	151.6363	1.281	151.7534	1.681	152.6264	1.055	792.9800
1.063	150.7114	1.076	151.6388	1.268	151.7558	1.735	152.6289	1.055	792.9887
1.062	150.7138	1.085	151.6412	1.275	151.7582	1.747	152.6313	1.055	792.9918
1.062	150.7162	1.076	151.6437	1.265	151.7607	1.770	152.6337	1.057	792.9934
1.063	150.7187	1.078	151.6461	1.269	151.7631	1.775	152.6362	1.066	792.9950
1.064	150.7211	1.080	151.6486	1.258	151.7655	1.765	152.6386	1.045	793.0010
1.073	150.7235	1.097	151.6510	1.253	151.7679	1.794	152.6411	1.081	793.0038
1.062	150.7260	1.085	151.6534	1.256	151.7704	1.785	152.6435	1.053	793.0054
1.075	150.7284	1.098	151.6559	1.225	151.7728	1.269	792.8506		
1.077	150.7308	1.091	151.6583	1.238	151.7752	1.202	792.8597		
1.073	150.7333	1.092	151.6607	1.223	151.7777	1.164	792.8641		
1.083	150.7357	1.093	151.6632	1.204	151.7801	1.160	792.8673		

Table 3. Residuals from the Linear and Quadratic period study of TYC 1488-693-1.

No.	Epochs	Cycles	Linear Residuals	Quadratic Residuals	Wt	Reference
1	51322.2596	−9783.0	−0.0087	−0.0038	0.2	NSVS (Pojmański 2013)
2	51336.2674	−9759.5	0.0052	0.0092	0.2	NSVS (Pojmański 2013)
3	51364.2454	−9712.5	−0.0044	−0.0024	0.2	NSVS (Pojmański 2013)
4	51375.2614	−9694.0	−0.0048	−0.0036	0.2	NSVS (Pojmański 2013)
5	51375.2624	−9694.0	−0.0038	−0.0026	0.2	NSVS (Pojmański 2013)
6	51599.4828	−9317.5	0.0173	0.0035	0.2	NSVS (Pojmański 2013)
7	57113.9330	−57.0	−0.0007	−0.0022	1.0	Present observations
8	57117.8039	−50.5	−0.0004	−0.0017	1.0	Present observations
9	57136.8600	−18.5	0.0002	0.0002	1.0	Present observations
10	57147.8761	0.0	−0.0002	0.0007	1.0	Present observations
11	57148.7704	1.5	0.0010	0.0019	1.0	Present observations
12	57151.7468	6.5	0.0000	0.0010	1.0	Present observations

Table 4. TYC 1488-693-1, light curve characteristics.

Filter	Phase	Magnitude Min. I	Phase	Magnitude Max. I
	0.00		0.25	
ΔB		1.21 ± 0.01		0.39 ± 0.01
ΔV		1.42 ± 0.01		0.64 ± 0.01
ΔR_c		1.62 ± 0.02		0.87 ± 0.01
ΔI_c		1.77 ± 0.01		1.06 ± 0.01

Filter	Phase	Magnitude Min. II	Phase	Magnitude Max. II
	0.50		0.75	
ΔB		0.56 ± 0.01		0.40 ± 0.01
ΔV		0.81 ± 0.01		0.63 ± 0.01
ΔR_c		1.06 ± 0.02		0.87 ± 0.01
ΔI_c		1.27 ± 0.01		1.06 ± 0.01

Filter	Min. I − Max. I	Max. I − Max. II
ΔB	0.81 ± 0.01	0.00 ± 0.01
ΔV	0.78 ± 0.02	0.01 ± 0.02
ΔR_c	0.75 ± 0.02	0.00 ± 0.01
ΔI_c	0.71 ± 0.02	0.00 ± 0.02

Filter	Min. I − Min. II	Min. II − Max. II
ΔB	0.65 ± 0.02	0.17 ± 0.02
ΔV	0.61 ± 0.02	0.18 ± 0.02
ΔR	0.56 ± 0.04	0.19 ± 0.01
ΔI	0.50 ± 0.02	0.22 ± 0.02

Table 5. TYC 1488-693-1, a light curve solution.

Parameters	Values
$\lambda_B, \lambda_V, \lambda_{Rc}, \lambda_{Ic}$ (nm)	440, 550, 640, 790
$x_{bol1,2}, y_{bol1,2}$	0.641, 0. 630, 0.232, 0.145
$x_{1Ic,2Ic}, y_{1Ic,2Ic}$	0. 569, 0.668, 0. 271, 0.144
$x_{1Rc,2Rc}, y_{1Rc,2Rc}$	0.652, 0.754, 0. 278, 0.096
$x_{1V,2V}, y1V,2V$	0.725, 0.799, 0. 266, 0.006
$x_{1B,2B}, y_{1B,2B}$	0. 815, 0.840, 0.206, −0.155
g_1, g_2	0.32
A_1, A_2	0.5
Inclination (°)	78.74 ± 0.04
T_1, T_2 (K)	6750, 4397 ± 2
Ω_1, Ω_2	3. 150 ± 0.001, 3.191 ± 0.002
$q(m_2/m_1)$	0.5829 ± 0.0007
Fill−outs: F_1, F_2	96.27 ± 0.04%, 95.03 ± 0.04%
$L_1/(L_1+L_2)_I$	0.8974 ± 0.0003
$L_1/(L_1+L_2)_R$	0.9215 ± 0.0003
$L_1/(L_1+L_2)_V$	0.9464 ± 0.0004
$L_1/(L_1+L_2)_B$	0.9720 ± 0.0003
JDo (days)	2457147.83765 ± 0.00024
Period (days)	0.5954652 ± 0. 0000015
r_1, r_2 (pole)	0. 3838 ± 0.0015, 0.288 ± 0.001
r1, r2 (point)	0.461 ± 0.004, 0.340 ± 0.003
r_1, r_2 (side)	0.403 ± 0.002, 0.298 ± 0.001
r_1, r_2 (back)	0.426 ± 0.002, 0.320 ± 0.002

Photometry and Light Curve Modeling of HO Piscium and V535 Pegasi

Rahim Heidarnia

RIAAM Observatory, Research Institute for Astronomy and Astrophysics of Maragha, Maragha, Iran; heidarnia.rahim@gmail.com

Received July 17, 2018; revised October 29, November 12, 2018; accepted November 12, 2018

Abstract   In this article we will present the photometric study of the overcontact binaries HO Psc and V535 Peg. The data were acquired with the 304-mm telescope of RIAAM Observatory, and after the data reduction and photometry, the main parameters of the systems such as temperatures, inclination, and mass ratio were found using modeling in PHOEBE software.

1. Introduction

Studies of eclipsing binary stars are currently of interest because of testing models and understanding their various intrinsic properties (Terrell 2006). HO Psc (Martignoni 2006) and V535 Peg (Geske *et al.* 2006) are also overcontact eclipsing binaries, so they share a common envelope of material. W UMa system light curves usually have equal depth of primary and secondary minima, that is because both components have almost equal temperature. In W UMa variables, usually the components are so close that gravitational effects causes deformations of components. Information on these stars can be seen in Figure 1 and Table 1.

2. Observation and data reduction

We used the Research Institute for Astronomy and Astrophysics of Maragheh (RIAAM) observatory equipment which include a 304.8mm schmidt-cassegrain telescope and a SBIG STX-16803 CCD. The data were captured from July 2016 to October 2017 in BVR filters. The telescope was guided with a DMK31AU03 CCD mounted on a small telescope with focal length of 1000 mm. We also used 2 × 2 binning, and the CCD's temperature was fixed on –35°C with 75% of cooling fan power. The IRAF package was used for reducing the bias and dark frames and also dividing the flat field frame, which we

captured in twilight from sky horizon at opposite direction of the sunrise. The reduced data were used for aperture photometry, so the photometry files were made in columns of HJD, object magnitude, and check star magnitude for each filter.

All of the magnitude points of the check star were subtracted from the average of them in order to find the variation range of the magnitude in observing time, and then they were subtracted from the object's magnitude (as seen in Tables 6 and 7). The HJD were converted to orbital phase using Equation 1 and the light curves have been plotted as seen in Figures 2 and 3.

$$\text{Phase} = \text{decimal}[(\text{HJD}_0 - \text{Epoch}) / \text{period}] \qquad (1)$$

The data of period of 0.4 to 0.6 and 0.9 to 1.1 for each minima were exported to an ASCII file in HJD and magnitude columns in order to calculate the minima time. The table curve software was used for peak fitting the data and calculating the minima time and error as seen in Table 2, and the O–C diagrams were plotted (Figure 4).

The data for HO Psc and V535 Peg are given in Tables 6 and 7, respectively, at the end of this article.

3. Light curve modelling

In order to find the physical parameters of the systems we tried to achieve the best fitted model using the PHOEBE software (Prša and Zwitter 2005) , which uses the Wilson-Devinney code (WD; Wilson and Devinney 1971). The data were imported as an ASCII file with columns of phase and magnitude of the objects. The initial values of the temperatures of the systems were used considering their color indexes. As we have close binaries, the difference between the surface temperatures of the components was almost equal.

The next important parameter in PHOEBE is the mass ratio (q) of the systems. We tried about 20 steps for initial value of the mass ratio from 0.4 to 3, and considering the shape of the synthetic light curve and the chi² value, q = 0.9 for HO Psc and q = 0.6 for V535 Peg were used.

The limb darkening values have been added from the van Hamme (1993) table with logarithmic law for the stars with

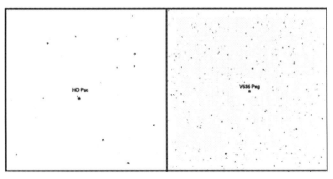

Figure 1. Field of view of the objects, left: HO Psc, right: V535 Peg.

Table 1. Objects.

Object	R.A. (2000) h m s	Dec. (2000) deg ' "	B	V	Magnitude (simbad) J	H	K
HO Psc	01 30 16.466	+13 33 25.08	11.0	11.50	9.659	9.29	9.215
V535 Peg	22 36 16.7640	+33 18 56.761	11.18	10.851	9.157	8.793	8.694

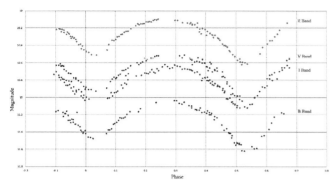

Figure 2. V535 Peg BVRI light curve.

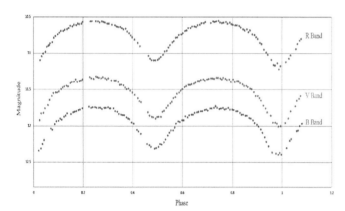

Figure 3. HO Psc BVR light curve.

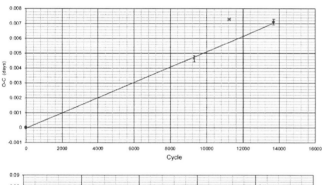

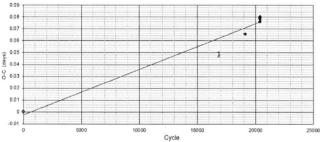

Figure 4. O–C diagrams for HO Psc (upper plot) and V535 Peg (lower plot).

$T_{eff} < 9000$ with convective layer (Al-Naimiy 1978). The gravity darkening values were also used from the table of Lucy (1967). After that, we tried to achieve the best physical parameters of inclination and temperature. With a lot of iterations, we used the calculation of PHOEBE to get the best fitted model considering the chi^2 without spot as seen in Table 3. In both systems' light curves, a difference in the magnitude was seen in the out-of-eclipse region that is known as the O'Connell effect (O'Connell

Table 2. Observed minima.

Object	Primary Minima	rms Error	Secondary Minima	rms Error
HO Psc	2457702.4888	4.1e^{-5}	2457702.3262	0.9e^{-5}
V535 Peg	2458035.3140	6.32e^{-5}	2458035.4725	1.83e^{-5}
	2458038.2224	1.22e^{-5}	2458038.383	3.91e^{-5}
	2458039.189	2.94e^{-5}	2458039.349	1.39e^{-5}

Table 3. Physical parameters.

Parameter	HO Psc	V535 Peg	Error
Period (days)	0.324747736	0.323003849	—
New epoch	2457702.4872	2458039.18664	—
Ω_1	3.35	2.85	0.03
Ω_2	3.35	2.85	0.03
q_{ptm}	0.90	0.58	0.01
Inclination	75.14°	72.56°	0.1
Limb Darkening (linear)	x1 = 0.68 y1 = 0.18	x1 = 0.67 y1 = 0.21	—
Limb Darkening (non-linear)	x2 = 0.68 y2 = 0.18	x2 = 0.67 y2 = 0.20	—
Gravity Darkening	g1 = 0.5	g1 = 0.32	—
	g2 = 0.82	g2 = 0.32	—
T_{eff1}	6674 K	6730 K	12
T_{eff2}	6228 K	6509 K	20
L_1 ($L_\odot$)	4.38	4.88	—
L_2 ($L_\odot$)	3.24	2.514	—
R_1 ($R_\odot$)	1.57	1.63	—
R_2 ($R_\odot$)	1.50	1.25	—
M_{bol1}	3.14	3.02	—
M_{bol2}	3.53	3.74	—
SMA	5.86	2.60	—

Table 4. HO Psc spot parameters.

	Colatitude	Longitude	Radius	Temperature
Primary Star	90	90	10	0.9

Table 5. V535 Peg spots parameters.

	Colatitude	Longitude	Radius	Temperature
Primary Star	90	90	20	0.7

1951), so we tried to add a spot (Tables 4 and 5). The fitted models of the systems are shown in Figures 5 and 6.

To test this model, we tried to calculate the luminosities and radii values using the emperical relationship between M_{bol} and T_{eff} given by Reed (1998) for the $T_{eff} < 9141$ as seen in Equation 2 and then, driving luminosity and radius with Equations 3 to 5.

$$BC = -8.499\,[\log(T) - 4]4 + 13.421[\log(T) - 4]3 - 8.131[\log(T) - 4]2 - 3.901\,[\log(T) - 4] - 0.438 \quad (2)$$

$$M_{bol} = M_v + BC(T_{eff}) \quad (3)$$

$$M_{bol(*)} = M_{bol(sun)} - 2.5\,\log(L * / L_{sun}) \quad (4)$$

$$R^2 = L / T_{eff}^4 \quad (5)$$

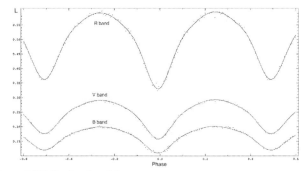

Figure 5. HO Psc fitted model.

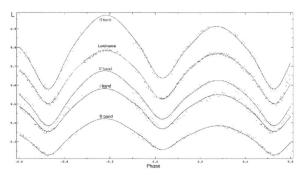

Figure 6. V535 Peg fitted model.

The measured values for radii and luminosities were the same as the values obtained in PHOEBE as seen in Table 3.

We also used the main sequence parameter table of Boyajian *et al.* (2013) for the mass of the components which were well matched with our stars, and considering the mass values we determined the semi-major axis values of the systems.

The 3D shapes of the systems were also drawn using BINARYMAKER software (Bradstreet and Steelman 2002; Figures 7 and 8).

4. Acknowledgement

This work has been supported financially by the Research Institute in Astronomy and Astrophysics of Maraghah.

References

Al-Naimiy, H. M. 1978, *Astrophys. Space Sci.*, **53**, 181.

Boyajian, T. S., *et al.* 2013, *Astrophys. J.*, **771**, 40.

Bradstreet, D. H., and Steelman, D. P. 2002, *Bull. Amer. Astron. Soc.*, **34**, 1224.

Geske M. T., Gettel S. J., and McKay T. A. 2006, *Astron. J.*, **131**, 633.

Lucy, L. B. 1967, *Z. Astrophys.*, **65**, 89.

Martignoni, M. 2006, *Inf. Bull. Var. Stars*, No. 5700, 1.

O'Connell, D. J. K. 1951, *Publ. Riverview Coll. Obs.*, **2**, 85.

Prša, A., and Zwitter, T. 2005, *Astrophys. J.*, **628**, 426.

Reed B. C. 1998, *J. Roy. Astron. Soc. Canada*, **92**, 36.

Terrel D. 2006, in *Astrophysics of Variable Stars*, eds. C. Sterken, C. Aerts, ASP Conf. Ser. 349, Astronomical Society of the Pacific, San Francisco, 91.

van Hamme. W. 1993, *Astron. J.*, **106**, 2096.

Wilson, R. E., and Devinney, E. J., 1971, *Astrophys. J.*, **166**, 605.

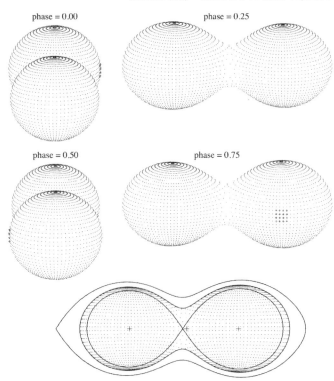

Figure 7. 3D shape of HO Psc.

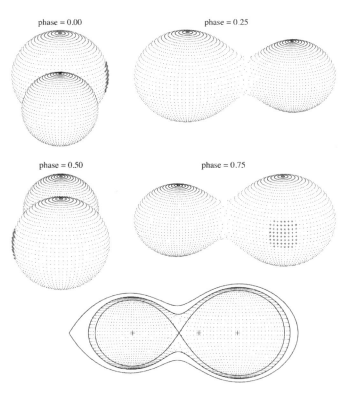

Figure 8. 3D shape of V535 Peg.

Table 6. HO Psc data.

HJD	B	HJD	V	HJD	R	HJD	B	HJD	V	HJD	R
2457702.172	12.34	2457702.173	11.92	2457702.174	11.10	2457702.350	12.02	2457702.346	11.67	2457702.344	10.89
2457702.175	12.31	2457702.175	11.83	2457702.176	11.04	2457702.352	11.97	2457702.348	11.62	2457702.347	10.85
2457702.177	12.25	2457702.178	11.82	2457702.178	11.00	2457702.355	11.94	2457702.351	11.60	2457702.349	10.84
2457702.179	12.20	2457702.180	11.76	2457702.181	10.98	2457702.357	11.95	2457702.353	11.56	2457702.352	10.79
2457702.188	12.04	2457702.184	11.71	2457702.183	10.92	2457702.360	11.92	2457702.356	11.54	2457702.354	10.78
2457702.190	12.05	2457702.186	11.64	2457702.185	10.90	2457702.362	11.92	2457702.358	11.52	2457702.356	10.75
2457702.193	11.98	2457702.189	11.59	2457702.186	10.89	2457702.364	11.87	2457702.360	11.50	2457702.359	10.74
2457702.195	11.94	2457702.191	11.55	2457702.187	10.87	2457702.367	11.88	2457702.363	11.47	2457702.361	10.72
2457702.198	11.93	2457702.194	11.54	2457702.190	10.85	2457702.369	11.85	2457702.365	11.45	2457702.364	10.70
2457702.200	11.90	2457702.196	11.54	2457702.192	10.79	2457702.372	11.84	2457702.368	11.43	2457702.366	10.69
2457702.202	11.91	2457702.198	11.53	2457702.194	10.76	2457702.374	11.84	2457702.370	11.43	2457702.368	10.66
2457702.205	11.88	2457702.201	11.49	2457702.197	10.75	2457702.376	11.82	2457702.372	11.42	2457702.371	10.66
2457702.207	11.86	2457702.203	11.47	2457702.199	10.72	2457702.379	11.82	2457702.375	11.42	2457702.373	10.64
2457702.209	11.83	2457702.205	11.46	2457702.201	10.69	2457702.381	11.80	2457702.377	11.41	2457702.376	10.64
2457702.214	11.84	2457702.208	11.45	2457702.204	10.70	2457702.384	11.80	2457702.380	11.39	2457702.378	10.64
2457702.217	11.82	2457702.210	11.46	2457702.206	10.68	2457702.386	11.80	2457702.382	11.39	2457702.380	10.62
2457702.219	11.81	2457702.213	11.41	2457702.209	10.67	2457702.389	11.78	2457702.384	11.37	2457702.383	10.61
2457702.221	11.80	2457702.215	11.40	2457702.211	10.66	2457702.391	11.77	2457702.387	11.36	2457702.385	10.59
2457702.224	11.79	2457702.218	11.39	2457702.213	10.63	2457702.393	11.79	2457702.389	11.38	2457702.388	10.59
2457702.226	11.78	2457702.220	11.37	2457702.216	10.63	2457702.396	11.76	2457702.392	11.36	2457702.390	10.57
2457702.229	11.76	2457702.222	11.39	2457702.218	10.61	2457702.398	11.76	2457702.394	11.35	2457702.393	10.62
2457702.238	11.74	2457702.225	11.38	2457702.221	10.62	2457702.401	11.75	2457702.397	11.35	2457702.395	10.63
2457702.241	11.75	2457702.227	11.38	2457702.223	10.59	2457702.403	11.73	2457702.399	11.35	2457702.397	10.59
2457702.243	11.74	2457702.229	11.37	2457702.225	10.60	2457702.405	11.75	2457702.401	11.36	2457702.400	10.57
2457702.246	11.75	2457702.239	11.33	2457702.228	10.60	2457702.408	11.77	2457702.404	11.34	2457702.402	10.57
2457702.248	11.75	2457702.241	11.35	2457702.230	10.57	2457702.410	11.76	2457702.406	11.34	2457702.404	10.57
2457702.251	11.75	2457702.244	11.36	2457702.240	10.56	2457702.413	11.75	2457702.409	11.35	2457702.407	10.56
2457702.253	11.75	2457702.247	11.33	2457702.242	10.56	2457702.415	11.78	2457702.411	11.37	2457702.409	10.58
2457702.255	11.77	2457702.249	11.34	2457702.244	10.60	2457702.417	11.77	2457702.413	11.34	2457702.412	10.59
2457702.258	11.75	2457702.251	11.34	2457702.247	10.56	2457702.420	11.75	2457702.416	11.36	2457702.414	10.58
2457702.260	11.75	2457702.254	11.35	2457702.250	10.56	2457702.422	11.77	2457702.418	11.37	2457702.417	10.58
2457702.263	11.75	2457702.256	11.34	2457702.252	10.58	2457702.425	11.78	2457702.421	11.37	2457702.419	10.60
2457702.265	11.76	2457702.259	11.35	2457702.255	10.57	2457702.427	11.78	2457702.423	11.39	2457702.422	10.58
2457702.266	11.76	2457702.261	11.36	2457702.257	10.57	2457702.430	11.80	2457702.426	11.37	2457702.424	10.60
2457702.270	11.76	2457702.263	11.35	2457702.259	10.58	2457702.432	11.80	2457702.428	11.39	2457702.426	10.61
2457702.272	11.80	2457702.267	11.38	2457702.262	10.59	2457702.435	11.82	2457702.431	11.39	2457702.429	10.62
2457702.275	11.81	2457702.271	11.38	2457702.264	10.59	2457702.437	11.84	2457702.433	11.39	2457702.431	10.62
2457702.277	11.84	2457702.273	11.38	2457702.268	10.61	2457702.439	11.84	2457702.435	11.41	2457702.434	10.62
2457702.280	11.84	2457702.276	11.40	2457702.272	10.62	2457702.442	11.86	2457702.438	11.43	2457702.436	10.65
2457702.282	11.85	2457702.278	11.43	2457702.274	10.64	2457702.444	11.86	2457702.440	11.45	2457702.439	10.67
2457702.284	11.85	2457702.280	11.44	2457702.276	10.64	2457702.447	11.89	2457702.443	11.45	2457702.441	10.69
2457702.287	11.89	2457702.283	11.44	2457702.279	10.64	2457702.449	11.90	2457702.445	11.47	2457702.444	10.68
2457702.289	11.90	2457702.285	11.47	2457702.281	10.65	2457702.452	11.92	2457702.448	11.49	2457702.446	10.71
2457702.291	11.91	2457702.287	11.49	2457702.283	10.67	2457702.454	11.96	2457702.450	11.50	2457702.448	10.69
2457702.294	11.95	2457702.290	11.49	2457702.286	10.66	2457702.457	11.98	2457702.453	11.53	2457702.451	10.74
2457702.297	11.97	2457702.292	11.52	2457702.288	10.70	2457702.459	12.01	2457702.455	11.59	2457702.454	10.75
2457702.301	12.05	2457702.295	11.54	2457702.291	10.73	2457702.462	12.05	2457702.458	11.59	2457702.456	10.78
2457702.303	12.08	2457702.298	11.57	2457702.293	10.76	2457702.464	12.10	2457702.460	11.62	2457702.458	10.80
2457702.306	12.11	2457702.302	11.62	2457702.295	10.78	2457702.467	12.16	2457702.462	11.65	2457702.461	10.86
2457702.314	12.25	2457702.304	11.66	2457702.298	10.81	2457702.469	12.22	2457702.465	11.72	2457702.463	10.88
2457702.317	12.26	2457702.306	11.67	2457702.302	10.86	2457702.472	12.28	2457702.468	11.76	2457702.466	10.95
2457702.319	12.29	2457702.315	11.82	2457702.305	10.91	2457702.477	12.37	2457702.470	11.81	2457702.469	10.99
2457702.321	12.32	2457702.317	11.86	2457702.316	11.04	2457702.480	12.31	2457702.475	11.86	2457702.471	11.02
2457702.324	12.31	2457702.320	11.88	2457702.318	11.08	2457702.482	12.38	2457702.478	11.95	2457702.476	11.09
2457702.326	12.30	2457702.322	11.89	2457702.320	11.10	2457702.485	12.40	2457702.481	11.96	2457702.479	11.16
2457702.329	12.29	2457702.325	11.90	2457702.323	11.10	2457702.487	12.38	2457702.483	11.97	2457702.481	11.17
2457702.331	12.25	2457702.327	11.87	2457702.325	11.10	2457702.489	12.39	2457702.485	12.01	2457702.484	11.14
2457702.333	12.26	2457702.329	11.88	2457702.328	11.09	2457702.496	12.30	2457702.488	12.00	2457702.486	11.23
2457702.336	12.23	2457702.332	11.87	2457702.330	11.07	2457702.500	12.20	2457702.494	11.96	2457702.488	11.18
2457702.338	12.19	2457702.334	11.81	2457702.332	11.06	2457702.502	12.18	2457702.498	11.90	2457702.495	11.12
2457702.341	12.13	2457702.336	11.79	2457702.335	11.02	2457702.509	12.07	2457702.501	11.80	2457702.499	11.08
2457702.343	12.10	2457702.339	11.75	2457702.337	10.98	2457702.511	12.07	2457702.507	11.73	2457702.501	11.03
2457702.345	12.06	2457702.341	11.72	2457702.340	10.97	2457702.513	12.00	2457702.509	11.65	2457702.508	10.94
2457702.348	12.06	2457702.344	11.69	2457702.342	10.92	2457702.516	11.97	2457702.512	11.63	2457702.510	10.88

Table 7. V535 Peg data.

HJD	B	HJD	V	HJD	R	HJD	B	HJD	V	HJD	R
2458038.175	11.52	2458038.176	10.58	2458038.176	10.60	2458039.147	11.45	2458039.153	10.59	2458039.151	10.46
2458038.178	11.50	2458038.178	10.57	2458038.179	10.60	2458039.148	11.30	2458039.155	10.51	2458039.153	10.59
2458038.181	11.53	2458038.181	10.56	2458038.182	10.63	2458039.150	11.33	2458039.156	10.61	2458039.155	10.64
2458038.183	11.56	2458038.184	10.49	2458038.184	10.64	2458039.152	11.56	2458039.158	10.64	2458039.157	10.67
2458038.186	11.60	2458038.187	10.64	2458038.187	10.62	2458039.154	11.59	2458039.160	10.66	2458039.159	10.68
2458038.189	11.56	2458038.189	10.65	2458038.190	10.63	2458039.156	11.55	2458039.162	10.75	2458039.161	10.69
2458038.191	11.61	2458038.192	10.67	2458038.192	10.70	2458039.158	11.62	2458039.164	10.78	2458039.162	10.71
2458038.194	11.66	2458038.195	10.71	2458038.195	10.71	2458039.160	11.63	2458039.166	10.74	2458039.164	10.75
2458038.197	11.66	2458038.197	10.71	2458038.198	10.75	2458039.162	11.61	2458039.168	10.77	2458039.166	10.76
2458038.199	11.69	2458038.200	10.77	2458038.200	10.80	2458039.163	11.61	2458039.171	10.79	2458039.168	10.84
2458038.202	11.75	2458038.203	10.77	2458038.203	10.82	2458039.165	11.68	2458039.173	10.73	2458039.171	10.79
2458038.205	11.78	2458038.205	10.81	2458038.206	10.87	2458039.167	11.60	2458039.175	10.85	2458039.173	10.84
2458038.207	11.78	2458038.208	10.83	2458038.208	10.87	2458039.171	11.73	2458039.177	10.82	2458039.175	10.82
2458038.210	11.81	2458038.211	10.87	2458038.211	10.89	2458039.172	11.76	2458039.199	10.81	2458039.200	10.84
2458038.213	11.82	2458038.213	10.89	2458038.214	10.91	2458039.174	11.76	2458039.201	10.76	2458039.202	10.76
2458038.216	11.83	2458038.219	10.86	2458038.219	10.92	2458039.176	11.73	2458039.203	10.76	2458039.204	10.79
2458038.218	11.96	2458038.253	10.62	2458038.254	10.63	2458039.199	11.77	2458039.205	10.72	2458039.205	10.70
2458038.253	11.56	2458038.256	10.55	2458038.257	10.59	2458039.201	11.74	2458039.207	10.69	2458039.207	10.75
2458038.256	11.50	2458038.293	10.44	2458038.293	10.52	2458039.203	11.75	2458039.209	10.71	2458039.209	10.69
2458038.292	11.44	2458038.296	10.46	2458038.296	10.53	2458039.204	11.72	2458039.211	10.68	2458039.211	10.72
2458038.295	11.43	2458038.298	10.43	2458038.299	10.49	2458039.206	11.68	2458039.212	10.68	2458039.213	10.71
2458038.298	11.42	2458038.301	10.46	2458038.301	10.44	2458039.208	11.67	2458039.214	10.69	2458039.215	10.78
2458038.300	11.42	2458038.303	10.50	2458038.304	10.52	2458039.210	11.66	2458039.216	10.64	2458039.217	10.63
2458038.303	11.38	2458038.306	10.49	2458038.307	10.54	2458039.212	11.63	2458039.218	10.59	2458039.218	10.63
2458038.306	11.35	2458038.309	10.47	2458038.309	10.54	2458039.214	11.59	2458039.220	10.60	2458039.220	10.66
2458038.308	11.41	2458038.311	10.44	2458038.312	10.52	2458039.216	11.62	2458039.222	10.60	2458039.222	10.63
2458038.311	11.43	2458038.314	10.42	2458038.314	10.51	2458039.218	11.58	2458039.230	10.58	2458039.231	10.61
2458038.314	11.43	2458038.317	10.51	2458038.317	10.58	2458039.219	11.58	2458039.232	10.53	2458039.233	10.59
2458038.316	11.39	2458038.319	10.51	2458038.320	10.54	2458039.221	11.57	2458039.234	10.57	2458039.235	10.56
2458038.319	11.42	2458038.322	10.51	2458038.322	10.54	2458039.230	11.50	2458039.236	10.53	2458039.236	10.61
2458038.322	11.44	2458038.325	10.55	2458038.325	10.56	2458039.232	11.50	2458039.238	10.54	2458039.238	10.55
2458038.324	11.48	2458038.327	10.56	2458038.328	10.59	2458039.234	11.53	2458039.240	10.50	2458039.240	10.55
2458038.327	11.45	2458038.330	10.56	2458038.330	10.58	2458039.236	11.50	2458039.242	10.50	2458039.242	10.53
2458038.329	11.53	2458038.333	10.56	2458038.333	10.59	2458039.237	11.45	2458039.244	10.49	2458039.244	10.53
2458038.332	11.52	2458038.335	10.59	2458038.336	10.63	2458039.239	11.70	2458039.245	10.48	2458039.246	10.50
2458038.335	11.50	2458038.338	10.59	2458038.338	10.61	2458039.241	11.45	2458039.247	10.49	2458039.248	10.54
2458038.338	11.53	2458038.341	10.63	2458038.341	10.64	2458039.243	11.45	2458039.249	10.51	2458039.249	10.53
2458038.340	11.54	2458038.343	10.62	2458038.344	10.70	2458039.245	11.43	2458039.251	10.46	2458039.251	10.47
2458038.343	11.60	2458038.346	10.64	2458038.346	10.67	2458039.247	11.45	2458039.253	10.47	2458039.253	10.50
2458038.345	11.61	2458038.349	10.67	2458038.349	10.69	2458039.249	11.42	2458039.255	10.45	2458039.255	10.50
2458038.348	11.59	2458038.351	10.70	2458038.352	10.72	2458039.250	11.43	2458039.289	10.49	2458039.290	10.59
2458038.351	11.61	2458038.354	10.71	2458038.354	10.77	2458039.252	11.45	2458039.291	10.55	2458039.291	10.58
2458038.353	11.71	2458038.357	10.75	2458038.357	10.77	2458039.254	11.39	2458039.293	10.61	2458039.293	10.67
2458038.356	11.73	2458038.359	10.80	2458038.360	10.82	2458039.289	11.44	2458039.295	10.58	2458039.295	10.70
2458038.359	11.73	2458038.362	10.87	2458038.362	10.87	2458039.291	11.45	2458039.297	10.52	2458039.297	10.61
2458038.361	11.76	2458038.365	10.86	2458038.365	10.90	2458039.292	11.52	2458039.299	10.56	2458039.299	10.64
2458038.364	11.83	2458038.367	10.90	2458038.368	10.96	2458039.294	11.50	2458039.300	10.65	2458039.301	10.63
2458038.367	11.85	2458038.370	10.95	2458038.370	10.96	2458039.296	11.52	2458039.302	10.52	2458039.303	10.61
2458038.369	11.93	2458038.373	10.96	2458038.373	10.99	2458039.298	11.48	2458039.304	10.55	2458039.305	10.72
2458038.372	11.89	2458038.375	10.96	2458038.376	11.02	2458039.300	11.46	2458039.306	10.57	2458039.306	10.67
2458038.375	11.95	2458038.378	11.04	2458038.378	11.01	2458039.302	11.49	2458039.308	10.64	2458039.308	10.66
2458038.377	11.91	2458038.381	10.97	2458038.381	11.02	2458039.304	11.52	2458039.310	10.60	2458039.310	10.64
2458038.380	11.97	2458038.383	11.03	2458038.384	11.00	2458039.306	11.54	2458039.312	10.54	2458039.312	10.67
2458038.383	11.95	2458038.386	10.97	2458038.386	10.95	2458039.307	11.52	2458039.313	10.69	2458039.314	10.70
2458038.385	11.93	2458038.389	10.95	2458038.389	10.94	2458039.309	11.55	2458039.315	10.63	2458039.316	10.70
2458038.388	11.96	2458038.391	10.88	2458038.392	10.87	2458039.311	11.57	2458039.317	10.65	2458039.318	10.65
2458038.391	11.89	2458038.394	10.86	2458038.394	10.89	2458039.313	11.63	2458039.319	10.73	2458039.319	10.72
2458038.393	11.83	2458038.397	10.82	2458038.397	10.87	2458039.315	11.58	2458039.321	10.78	2458039.321	10.71
2458038.396	11.75	2458038.399	10.71	2458038.400	10.76	2458039.317	11.59	2458039.323	10.66	2458039.323	10.85
2458038.399	11.78	2458038.402	10.75	2458038.402	10.77	2458039.319	11.60	2458039.325	10.69	2458039.325	10.77
2458038.401	11.69	2458038.405	10.65	2458038.405	10.71	2458039.320	11.60	2458039.326	10.78	2458039.327	10.81
2458038.404	11.70	2458038.407	10.67	2458038.408	10.71	2458039.322	11.72	2458039.328	10.78	2458039.329	10.80
2458038.407	11.55	2458038.410	10.64	2458038.410	10.66	2458039.324	11.73	2458039.330	10.86	2458039.331	10.84
2458038.409	11.53	2458038.413	10.54	2458038.413	10.60	2458039.326	11.72	2458039.332	10.88	2458039.332	10.88
2458038.412	11.50	2458038.415	10.62	2458038.416	10.59	2458039.328	11.73	2458039.334	10.88	2458039.334	10.93
2458038.415	11.46	2458038.418	10.56	2458038.418	10.63	2458039.330	11.76	2458039.336	10.91	2458039.336	10.90
2458038.417	11.44	2458038.421	10.50	2458038.421	10.56	2458039.332	11.79	2458039.338	10.94	2458039.338	10.95
2458038.420	11.47	2458038.423	10.44	2458038.424	10.52	2458039.333	11.79	2458039.339	10.92	2458039.340	10.95
2458038.423	11.41	2458039.147	10.45	2458038.426	10.49	2458039.335	11.83	2458039.341	10.99	2458039.342	10.96
2458038.425	11.27	2458039.149	10.32	2458039.148	10.34	2458039.337	11.92	2458039.343	11.04	2458039.344	11.02
2458038.428	11.25	2458039.151	10.35	2458039.149	10.35						

A Study of Pulsation and Fadings in some R Coronae Borealis (RCB) Stars

John R. Percy

Department of Astronomy and Astrophysics, and Dunlap Institute of Astronomy and Astrophysics, University of Toronto, 50 St. George Street, Toronto, ON M5S 3H4, Canada; john.percy@utoronto.ca

Kevin H. Dembski

Department of Astronomy and Astrophysics, University of Toronto, 50 St. George Street, Toronto, ON, M5S 3H4, Canada; kevin.dembski@mail.utoronto.ca

Received, September 10, 2018; revised October 16 2018; accepted November 5, 2018

Abstract   We have measured the times of onset of recent fadings in four R Coronae Borealis (RCB) stars—V854 Cen, RY Sgr, R CrB, and S Aps. These times continue to be locked to the stars' pulsation periods, though with some scatter. In RY Sgr, the onsets of fading tend to occur at or a few days after pulsation maximum. We have studied the pulsation properties of RY Sgr through its recent long maximum using (O–C) analysis and wavelet analysis. The period "wanders" by a few percent. This wandering can be modelled by random cycle-to–cycle period fluctuations, as in some other types of pulsating stars. The pulsation amplitude varies between 0.05 and 0.25 in visual light, non-periodically but on a time scale of about 20 pulsation periods.

1. Introduction

R Coronae Borealis (RCB) stars are rare carbon-rich, hydrogen-poor, highly-evolved yellow supergiants which undergo fadings of up to 10 magnitudes, then slowly return to normal (maximum) brightness; see Clayton (2012) for an excellent review. Most or all RCB stars also undergo small-amplitude pulsations with periods of a few weeks. Although it was once considered that the fadings were random, it is now known that, in at least some RCB stars, the fadings are locked to the pulsation period, i. e., the onsets of the fadings occur at about the same phase of the pulsation cycle (Pugach 1977; Lawson *et al.* 1992; Crause *et al.* 2007, hereinafter CLH). This suggests a causal connection: e. g., the pulsation ejects a cloud of gas and dust; when this cools, the carbon condenses into soot; if the cloud lies between the observer and the star, the star appears to fade; it slowly reappears as the cloud disperses. It is also possible that temperature and density fluctuations in the stellar atmosphere, during the pulsations, lead to dust condensation (e. g., Woitke *et al.* 1996). Either case implies that the ejection is not radially symmetric; a cloud is ejected, not a shell.

Our interest in these stars was sparked by a somewhat-accidental encounter with the RCB star Z UMi (Percy and Qiu 2018). We had been studying Mira stars, and Z UMi had been misclassified as a Mira star in the *General Catalogue of Variable Stars* (GCVS; Samus *et al.* 2017) and in VSX (Watson *et al.* 2014), even though it had been identified as an RCB star by Benson *et al.* 1994). This star did not have a definitive pulsation period, but we measured the times of onset of its fadings, and we found that they were "locked" to a period of 41. 98 days, a typical pulsation period for an RCB star.

2. Data and analysis

We used visual observations from the AAVSO International Database (AID; Kafka 2018), the AAVSO vstar time-series analysis package (Benn 2013) which includes Fourier analysis, wavelet analysis, and polynomial fitting routines, and (O–C)

analysis to study the five RCB stars previously studied by CLH, and to study the pulsation of RY Sgr in more detail.

3. Results

3.1. Pulsation-fading relationships in RCB stars

CLH showed that, in five RCB stars, the times of onset of fadings were locked to their pulsation periods. The five stars, and their pulsation periods in days, were: V854 Cen (43. 25), RY Sgr (37. 79), UW Cen (42. 79), R CrB (42. 97), and S Aps (42. 99).

In the first part of this project, we determined the times of onset of fadings of these five stars since the work of CLH. The determination of these times was non-trivial. The visual observations had a typical uncertainty of 0.2 magnitude. Sometimes the data were sparse, and the exact time of onset was not well covered by the observations. This is especially true if the onsets fell within the seasonal gaps in the stars' observations. Some onsets could therefore not be measured. To determine the times, we experimented with fitting horizontal lines to the light curves preceding fadings, and sloping lines to the light curves following the onset of fadings, as well as using "by-eye" judgment.

We first determined, independently, the times published by CLH. Our times differed on average by ±4 days, which is the typical uncertainty of our determinations and CLH's. The differences averaged only ±3 days for R CrB, presumably the most densely-observed star. On average, our times were +1 day later than CLH's, which is not significantly different from zero. Our times are given in Table 1. The ephemerides are the same as used by CLH. UW Cen did not have any recent fadings whose times could be determined. The first time listed for each star is our redetermination of the time of onset of the last fading observed by CLH. This is followed by the CLH determination. This provides an indication of the difference and uncertainty in the timings. Times are labelled with a colon (:) if there was some scatter and/or sparseness in the data, or with a double colon (::) if there was much scatter and/or sparseness.

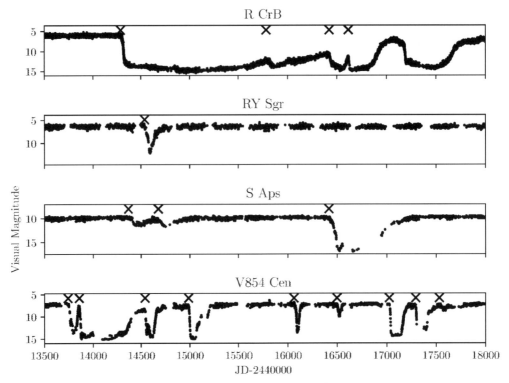

Figure 1. The recent AAVSO visual light curves of R CrB, RY Sgr, S Aps, and V854 Cen. The times of fadings (Table 1), as measured by us, are marked with an ×.

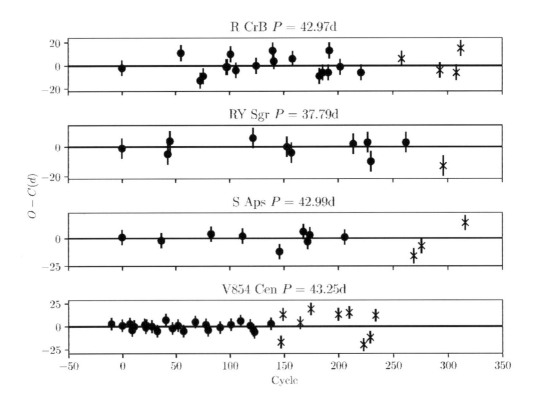

Figure 2. The (O–C) diagrams for the times of onsets of fadings in R CrB, RY Sgr, S Aps, and V854 Cen, using the periods given in section 3.1, and the observed times and cycle numbers listed in Table 1. The filled circles are the (O–C)s published by CLH.

Figure 1 shows the light curves, using the same format as CLH. The times of onset of fadings are indicated with an ×. Figure 2 shows the (O–C)s between our times of onset of fadings, and the pulsation ephemerides used by CLH. The average (O–C) for our times is almost twice that for CLH's times. This will be discussed in section 4.

3.2. Times of pulsation maximum in RY Sgr

RY Sgr has the largest pulsation amplitude of any known RCB star, though it is only about 0.15 in V. Several groups have observed or discussed the pulsation of RY Sgr for the purpose of determining and interpreting its apparent period change: Kilkenny (1982), Lawson and Cottrell (1990), Lombard and Koen (1993), Menzies and Feast (1997), among others.

RY Sgr has been at maximum since JD 2454900. In order to investigate the pulsation period, we have determined the times of 58 pulsation maxima between JD 2455031 and JD 2458243. They are listed in Table 2. They were determined independently by both of us, using low-order polynomial fitting (KHD, JRP) and phase-curve fitting (JRP), and then appropriately averaged.

3.3. Pulsation period variations in RY Sgr

The authors who were mentioned in section 3.2 determined the apparent period change in RY Sgr, and suggested various interpretations, including smoothly-varying period changes, and abrupt period changes. We have used two methods to investigate the period change: (O–C) analysis, and wavelet analysis, and applied them to the times in Table 2.

Figure 3 shows the (O–C) diagram for RY Sgr, using the times of maximum listed in Table 2, and a period of 37. 91 days. The scatter is consistent with the uncertainties in the times of maximum. Figure 4 (top) shows the period variation determined by wavelet analysis, using the WWZ routine in vstar. Both figures show that the period "wanders" between values of 37.0 and 38.5 days, with the period being approximately constant in the first third of the interval, increasing to a higher value in the second third, and decreasing to a lower value in the final third. The variation is not periodic, but its time scale is about 20 pulsation periods. In Mira stars, the time scale averages about 40 pulsation periods (Percy and Qiu 2018).

3.4. Are the period variations due to random cycle-to-cycle fluctuations?

The "wandering" pulsation periods of large-amplitude pulsating red giant stars (Mira stars) have been modelled by random cycle-to-cycle period fluctuations (Eddington and Plakidis 1929; Percy and Colivas 1999). We have investigated whether the period variations in RY Sgr can be modelled in this way by applying the Eddington-Plakidis formalism to the times of pulsation maximum given in Table 2. For this, we used a program written by one of us (KHD) in Python. We first tested it (successfully) on times of maximum of Mira, for comparison with Figure 1 in Percy and Colivas (1999).

In Figure 3, we showed the (O–C) values for RY Sgr, using a period of 37.91 days. Then, following Eddington and Plakidis (1929): let a(r) be the (O–C) of the rth maximum, and let ux(r) = a(r + x) – a(r), and $\overline{ux^2}$ be the average value, without regard to sign, of ux(r) for as many values of r as the observational

Table 1. New times of onset of fadings in four RCB stars.

Star	Cycle (n)	JD (obs)	JD (calc)	O–C (d)	Note
S Aps	206	2451670	2451674	–4	PD
—	206	2451675	2451674	1	CLH
—	269	2454366::	2454382	–16	—
—	276	2454676:	2454683	–7	—
—	316	2456417	2456403	14	—
RY Sgr	262	2453273	2453266	7	PD
—	262	2453269	2453266	3	CLH
—	296	2454538:	2454551	–13	—
V854 Cen	138	2453376	2453368	8	PD
—	138	2453371	2453368	3	CLH
—	147	2453740::	2453757	–17	—
—	149	2453856	2453843	13	—
—	165	2454540:	2454536	4	—
—	175	2454987	2454968	19	—
—	200	2456062	2456049	13	—
—	210	2456497	2456482	15	—
—	223	2457024	2457044	–20	—
—	229	2457292:	2457304	–12	—
—	234	2457532	2457520	12	—
R CrB	221	2452678	2452689	–11	PD
—	221	2452683	2452689	–6	CLH
—	258	2454285	2454279	6	—
—	293	2455779	2455783	–4	—
—	308	2456421	2456427	–6	—
—	312	2456615	2456599	15	—

Table 2. Times of pulsation maximum in RY Sgr (JD – 2400000).

JD (max)	JD (max)	JD (max)	JD (max)
55031	55740	56466	57274
55064	55787	56509	57309
55104	55823	56548	57600
55143	55863	56582	57653
55258	56018	56618	57683
55301	56046	56774	57721
55332	56085	56812	57906
55367	56125	56847	57940
55409	56167	56891	57984
55448	56197	56924	58018
55486	56239	56958	58055
55520	56268	57113	58205
55629	56349	57155	58243
55673	56394	57193	
55714	56427	57231	

material admits, then $\overline{ux^2} = 2a^2 + xe^2$ where a is the average observational error in determining the time of maximum, and e the average fluctuation in period, per cycle. A graph of $\overline{ux^2}$ versus x (the "Eddington-Plakidis diagram") should be a straight line if random cycle-to-cycle fluctuations occur. Figure 5 shows the $\overline{ux^2}$ versus x graph for RY Sgr, using the times listed in Table 2. The graph is approximately linear, with scatter which is not unexpected, given the limitations of the data. The value of a = 2.7 days is consistent with the errors in the measured times of maximum.

We also generated a $\overline{ux^2}$ versus x graph for RY Sgr, using the times of pulsation maximum published by Lawson and Cottrell (1990). They extend from JD 2441753 to 2447642. The graph is shown in Figure 6. The graph is approximately linear. The slope is comparable with that in Figure 5, and the value of a = 3.1 days is consistent with the expected errors in the measured times of maximum. The slopes e are 0.9 and 1.0 day for our data and Lawson and Cottrell's, respectively.

3.5. Pulsation amplitude variations in RY Sgr

Most of the famous Cepheid pulsating variables have constant pulsation amplitudes, but this is not true of other types, especially low-gravity stars: pulsating red giants (Percy and Abachi 2013), pulsating red supergiants (Percy and Khatu 2014), and some pulsating yellow supergiants (Percy and Kim 2014). The amplitudes of these stars vary by up to a factor of ten, on time scales of 20–30 pulsation periods.

We have used wavelet analysis to determine the variation in pulsation amplitude in RY Sgr, during the last 3,000 days when the star was at maximum (JD 2455000 to JD 2458250). The results are shown in the lower panel in Figure 4. The visual amplitude varies between 0.05 and 0.25. The amplitude variations can be confirmed by Fourier analysis of subsets of the data. The variation is not periodic, but occurs on a time scale of about 20 pulsation periods. This time scale is comparable to that found in the pulsating star types mentioned above. There is no strong consistency to the direction of the changes, though there is a slight tendency for the amplitude to be relatively medium-to-high at the beginning of a maximum, and relatively medium-to-low at the end.

We used the same method to study the amplitude variations during several shorter intervals when the star was at maximum. The results are given in Table 3. During these intervals, the visual amplitude also varies between 0.05 and 0.25. The median time scale of visual amplitude variation is about 30 (range 15 to 55) pulsation periods.

3.6. At what pulsation phase does the onset of fading occur?

CLH stated "...the absolute phase of the decline onsets could not be determined from the AAVSO data...." We have attempted to make this determination for RY Sgr, as follows. For each of the times of onset of fading determined by CLH, we have examined the previous 50–60 days of data, and measured the times of pulsation maximum using the same methods as in section 3.2. The results are listed in Table 4. The times of onset of fadings are the predicted times, given by CLH. There is considerable scatter, as there was in measuring the times of maximum in Table 2. Before some fadings, the data were too sparse to measure the pulsation maximum.

On average, the onsets of fading occur 7 days after pulsation maximum. According to Pugach (1977), the onset of fadings occurs at pulsation maximum, and the same is true for V854 Cen (Lawson *et al.* 1992). These conclusions depend, to some extent, on the definition of when the onset occurs, and may not be in conflict.

4. Discussion

The times of onset of fadings that we have measured (Table 1, Figure 1) seem to continue to be locked to the pulsation periods (Figure 2), though with a scatter ±11 days which is twice that obtained by CLH, even though our measured times are consistent with theirs. There are several possible explanations: (1) our times are actually less accurate than theirs;

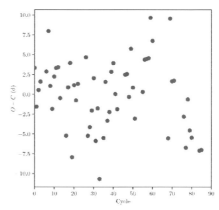

Figure 3. The (O–C) diagram for the times of pulsation maximum of RY Sgr listed in Table 2, using a period of 37.91 days. The period is approximately constant through cycles 0–25, slightly larger than average (upward slope) through cycles 25–60, and slightly smaller than average (downward slope) through cycles 60–90.

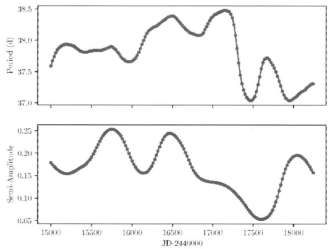

Figure 4. The variation in the pulsation period (top) and amplitude (bottom) of RY Sgr versus time, determined using the WWZ wavelet routine in vstar, and AAVSO visual observations.

Table 3. Pulsation amplitude variations in RY Sgr.

JD Range	Amplitude Range
2432950–2435437	0.06–0.20
2436597–2438054	0.14–0.20
2438303–2439653	0.07–0.23
2442095–2443300	0.17–0.11
2445739–2447949	0.07–0.20
2449886–2451403	0.10–0.24
2452078–2453227	0.04–0.15

Table 4. Times of onset of fading and of pulsation maximum in RY Sgr.

Onset of Fading F	Pulsation Maximum M	F–M (d)
2443366	2443357	+9
2444990	2444970	+20
2445066	2445064	+2
2447976	2447977	–1
2449147	2449133	+14
2441452	2441458	–6
2452057	2452046	+11
2453266	2453260	+6

(2) the "wandering" period (Figures 3 and 4) causes some scatter; (3) the fadings are not exactly locked to the pulsation; there are random factors in the pulsation, mass ejection, and onset of fadings which add to the scatter; (4) the differences are a statistical anomaly. We recognize that it is challenging to determine times of onset of fadings, or times of pulsation maxima, using visual data. This is where the density of the visual data can often help.

The interpretation and misinterpretation of period changes, especially as determined from (O–C) diagrams, has a long history, and a whole conference was devoted to this topic (Sterken 2005). If the (O–C) diagram has the appearance of a broken straight line, even with much scatter, it is often interpreted as an abrupt period change, e. g., Lawson and Cottrell (1990). If the (O–C) diagram is curved, even with much scatter, it is often interpreted as a smooth evolutionary change, e.g., Kilkenny (1982). Hundreds of (O–C) diagrams of Mira stars show these and other appearances, and can be modelled as due to random, cycle-to-cycle period fluctuations (Eddington and Plakidis 1929; Percy and Colivas 1999).

Our results (the linearity of Figures 5 and 6) suggest that the period variations in RY Sgr can be modelled, at least in part, by random cycle-to-cycle variations, as in Mira stars, rather than solely by a smooth evolutionary variation, or an abrupt variation. We cannot rule out the presence of a small smooth or abrupt variation but, if so, it is buried in the random period-fluctuation noise. The cause of the fluctuations is not known, but may be connected with the presence of large convective cells in the outer layers of the stars. The fact that the star ejects clouds, rather than shells, suggests that the outer layers of the star are not radially symmetric.

The discovery of a variable pulsation amplitude in RY Sgr (Figure 4) is an interesting but not-unexpected result, given the presence of amplitude variations in other low-gravity pulsating stars. Fernie (1989) and Lawson (1991) both pointed out that the pulsation amplitude of R CrB varied from cycle to cycle, but did not investigate the time scale of this phenomenon. We note that, when the pulsation amplitude is at its lowest, it is even more difficult to measure the times of pulsation maximum.

There is also the possibility that some RCB stars have two or more pulsation periods, either simultaneously or sequentially. Both Fernie (1989) and Lawson (1991) found a variety of periods in R CrB: Fernie (1989) found only 43.8 ± 0.1 days in 1985–1987, but 26.8, 44.4, and 73.7 days (possibly an alias) in 1972; Lawson (1991) found 51.8 and possibly 56.2 days in 1986–1989. Fernie (1989) considered that 26.8 and 44.4 days could possibly be the first overtone and fundamental periods, but 51.8 and 56.2 days are too close together to be radial overtones.

There are, unfortunately, many problems in determining these periods. The precision dP/P of periods P, determined from a single season of data, is limited to P/L, where L is the length of the dataset; see Figure 1 in Lawson (1991). If the period "wanders," the Fourier peaks will be further broadened. If the amplitude of the pulsation is changing, then Fourier analysis will give more than one period, whether or not these periods are real. If periods are determined from two or more seasons of data, then there will be alias periods; see Figure 2 in Fernie

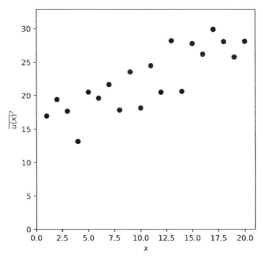

Figure 5. The Eddington-Plakidis diagram for RY Sgr, using the times of pulsation maximum in Table 2, and a period of 37.91 days.

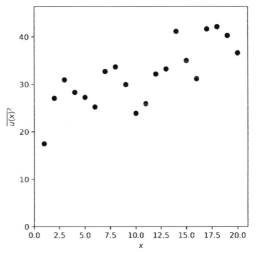

Figure 6. The Eddington-Plakidis diagram for RY Sgr, using the times of pulsation maximum given by Lawson and Cottrell (1990) between JD 2441753 and 2447642, and a period of 38.56 days.

(1989). There may be undetected variability due to minor dust obscuration, or other low-amplitude processes. For all these reasons, it is difficult to draw firm conclusions about multiple or variable pulsation periods.

There are other RCB stars which are known or suspected to pulsate. Rao and Lambert (2015) list 29. Most if not all of them have pulsation amplitudes which are even smaller than that of RY Sgr, so it will be almost impossible to study their pulsation with visual data. At least one of the 29 stars has an incorrect period: Z UMi is listed as having a period of 130 days, but Percy and Qiu (2018) were not able to fit AAVSO data to that period but, as mentioned in the Introduction, using the observed times of onset of fadings, they suggested a period of 41.98 days instead. It may be possible to determine times of onset of fadings of some of the 29 stars, and see if there is a period to which they are locked, as Percy and Qiu (2018) did for Z UMi.

In the future, most of these stars will be monitored through facilities such as LSST (the Large Synoptic Survey Telescope), but the century of archival AAVSO data will remain unique.

5. Conclusions

We have derived new information about the pulsation of the RCB star RY Sgr, especially about the variation of its period and amplitude. We have also strengthened the connection between the pulsation and the fadings in this star. We have used long-term archival visual data but, since the pulsation amplitudes of other RCB stars are even smaller than that of RY Sgr, future studies like ours will have to use long-term precision photoelectric or CCD observations.

6. Acknowledgements

We thank the AAVSO observers who made the observations on which this project is based, the AAVSO staff who archived them and made them publicly available, and the developers of the VSTAR package which we used for analysis. Coauthor KHD was a participant in the University of Toronto Work-Study Program, which we thank for administrative and financial support.

This project made use of the SIMBAD database, maintained in Strasbourg, France. The Dunlap Institute is funded through an endowment established by the David Dunlap family and the University of Toronto.

References

Benn, D. 2013, VSTAR data analysis software (http://www. aavso. org/vstar-overview).

Benson, P. J., Clayton, G. C., Garnavich, P., and Szkody, P. 1994, *Astron. J.*, **108**, 247.

Clayton, G. C. 2012, *J. Amer. Assoc. Var. Star Obs.*, **40**, 539.

Crause, L. A., Lawson, W. A., and Henden, A. A. 2007, *Mon. Not. Roy. Astron. Soc.*, **375**, 301.

Eddington, A. S., and Plakidis, S. 1929, *Mon. Not. Roy. Astron. Soc.*, **90**, 65.

Fernie, J. D. 1989, *Publ. Astron. Soc. Pacific*, **101**, 166.

Kafka, S. 2018, variable star observations from the AAVSO International Database (https://www. aavso. org/aavso-international-database)

Kilkenny, D. 1982, *Mon. Not. Roy. Astron. Soc.*, **200**, 1019.

Lawson, W. A. 1991, *Mon. Not. Royal Astron. Soc.*, **253**, 625.

Lawson, W. A., and Cottrell, P. L. 1990, *Mon. Not. Roy. Astron. Soc.*, **242**, 259.

Lawson, W. A., Cottrell, P. L., Gilmore, A. C., and Kilmartin, P. M. 1992, *Mon. Not. Roy. Astron. Soc.*, **256**, 339.

Lombard, F., and Koen, C. 1993, *Mon. Not. Roy. Astron. Soc.*, **263**, 309.

Menzies, J. W., and Feast, M. W. 1997, *Mon. Not. Roy. Astron. Soc.*, **285**, 358.

Percy, J. R., and Abachi, R. 2013, *J. Amer. Assoc. Var. Star Obs.*, **41**, 193.

Percy, J. R., and Colivas, T. 1999, *Publ. Astron. Soc. Pacific*, **111**, 94.

Percy, J. R., and Khatu, V. C. 2014, *J. Amer. Assoc. Var. Star Obs.*, **42**, 1.

Percy, J. R., and Kim, R. Y. H. 2014, *J. Amer. Assoc. Var. Star Obs.*, **42**, 267.

Percy, J. R., and Qiu, A. L. 2018, arxiv. org/abs/1805. 11027.

Pugach, A. F. 1977, *Inf. Bull. Var. Stars*, No. 1277, 1.

Rao, N. K., and Lambert, D. L. 2015, *Mon. Not. Roy. Astron. Soc.*, **447**, 3664.

Samus, N. N. *et al.*, 2017, *General Catalogue of Variable Stars*, Sternberg Astronomical Institute, Moscow (www. sai. msu. ru/gcvs/gcvs/index. htm).

Sterken, C. 2005, *The Light-Time Effect in Astrophysics: Causes and Cures of the (O–C) Diagram*, ASP Conf. Ser. 335, Astronomical Society of the Pacific, San Francisco.

Watson, C., Henden, A. A., and Price, C. A. 2014, AAVSO International Variable Star Index VSX (Watson+, 2006–2014; http://www.aavso.org/vsx).

Woitke, P., Goeres, A., and Sedlmayr, E. 1996, *Astron. Astrophys.*, **313**, 217.

Multi-color Photometry, Roche Lobe Analysis and Period Study of the Overcontact Binary System, GW Bootis

Kevin B. Alton

UnderOak Observatory, 70 Summit Ave, Cedar Knolls, NJ 07927; kbalton@optonline.net

Received September 11, 2018; revised October 23, 2018; accepted October 25, 2018

Abstract   GW Boo is a relatively bright (V-mag ~ 10.2) eclipsing W UMa binary system (P = 0.513544 d) which has surprisingly escaped detailed study since the first monochromatic light curve (LC) was published in 2003. LC data collected in 2011 and 2017 (B, V and I$_c$) at UnderOak Observatory (UO), produced eight new times-of-minimum for GW Boo which were used along with other eclipse timings from the literature to update the linear ephemeris. Secular variations (P$_3$ ~ 10.5y) in the orbital period suggested the possibility of a gravitationally bound third body. Roche modeling to produce synthetic LC fits to the observed data was accomplished using PHOEBE 0.31a and WDWINT 56a. In order to achieve the best synthetic fits to the multi-color LCs collected in 2011 and 2017 cool spot(s) were added to the Roche model.

1. Introduction

The variable behavior of GW Boo (GSC 1473-1049; BD+20°2890) was initially observed from data collected during the Semi-Automatic Variability Search (SAVS) (Maciejewski *et al.* 2003). Sparsely sampled photometric data for this system are available from the ROTSE-I survey (Akerlof *et al.* 2000; Wozniak *et al.* 2004; Gettel *et al.* 2006) as well as the ASAS survey (Pojmański *et al.* 2005). Although other times-of-minimum light have been sporadically published since 2008, this paper marks the first detailed period analysis and multi-color Roche model assessment of LCs for this system in the literature.

2. Observations and data reduction

Photometric collection dates at UnderOak Observatory (UO) included eight sessions between 03 June 2011 and 07 July 2011 with an additional 11-day imaging campaign conducted from 08 June 2017 to 26 June 2017. Instruments included a 0.2-m catadioptic telescope coupled with an SBIG ST-402ME CCD camera (2011) and a 0.28-m Schmidt-Cassegrain telescope (2017) equipped with an SBIG ST8-XME CCD camera; both were mounted at the Cassegrain focus. Automated imaging was performed with photometric B, V, and I$_c$ filters sourced from SBIG and manufactured to match the Bessell prescription; the exposure time for all dark- and light-frames was 60 seconds in 2011 and 75 seconds in 2017. As is standard practice at UO, the computer clock was automatically synchronized to a reference clock immediately prior to each session. Image acquisition (lights, darks, and flats) was performed using CCDSOFT v5 (Software Bisque 2011) or THESKYX Pro Version 10.5.0 (Software Bisque 2018) while calibration and registration were performed with AIP4WIN v2.4.0 (Berry and Burnell 2005). Images of GW Boo were plate solved using the standard star fields (MPOSC3) provided in MPO CANOPUS v10.7.1.3 (Minor Planet Observer 2015) in order to obtain the magnitude (B, V, and I$_c$ assignments for each comparison star. Only images taken above 30° altitude (airmass < 2.0) were accepted in order to minimize the effects of differential refraction and color extinction.

3. Results and discussion

3.1. Photometry and ephemerides

Five stars in the same field-of-view with GW Boo were used to derive catalog-based (MPOSC3) magnitudes in MPO CANOPUS (Table 1) using ensemble aperture photometry. During each imaging session comparison stars typically stayed within ± 0.011 mag for V and I$_c$ filters and ± 0.016 mag for B passband.

A total of 397 photometric values in B, 421 in V, and 426 in I$_c$ were acquired between 03 June 2011 and 07 July 2011 (Figure 1). The most recent campaign (08 June 2017– 28 June 2017) produced 409 values in B, 395 in V, and 414 in I$_c$ (Figure 2). Times-of-minimum were calculated using the method of Kwee and van Woerden (1956) as implemented in PERANSO v2.5 (Paunzen and Vanmunster 2016). Included in these determinations were eight new times-of-minimum for each filter which were averaged (Table 2) from each session. The Fourier routine (FALC; Harris *et al.* 1989) in MPO CANOPUS produced similar LC period solutions (0.531544 ± 0.000001 d) from both epochs. As appropriate, sparsely sampled photometric data from the ROTSE-I (clear filter) and ASAS (V-mag) surveys were converted from MJD to HJD and then normalized relative to V-mag data collected at UO in 2017. Period determinations from survey data were individually made from these data using PERANSO v2.5. The selected analysis method employed periodic orthogonal polynomials (Schwarzenberg-Czerny 1996) to fit observations and analysis of variance (ANOVA) to evaluate fit quality. The resulting orbital periods (P = 0.531544 ± 0.000008 d) were nearly identical and the folded curves remarkably superimposable (Figure 3). This provided an ideal opportunity to interpolate additional times-of-minimum from the survey data. A total of four values, two from each survey, that were closest to a mid-point bisecting line during Min I and Min II were weighted (50%) relative to directly observed new minima acquired at UO and published values (Table 2). These were used to analyze eclipse timings from 1999 through 2017 in which the reference epoch (Kreiner 2004) employed for calculating eclipse timing differences (ETD) was defined by the following linear ephemeris (Equation 1):

$$\text{Min.I(HJD)} = 2452500.335 + 0.5315444 \text{ E}. \qquad (1)$$

Table 1. Astrometric coordinates (J2000) and color indices (B-V) for GW Boo and five comparison stars used in this photometric study.

Star Identification	R.A. (J2000) h m s	Dec. (J2000) ° ' "	V-mag[a]	(B–V)[a]
GW Boo	13 53 13.85	+20 09 43.19	10.20	0.443
TYC 1473-1027-1	13 53 08.59	+20 07 24.00	11.38	0.478
GSC 1473-1036	13 53 19.14	+20 12 43.49	11.92	0.482
GSC 1473-0037	13 53 46.71	+20 11 17.99	12.85	0.258
TYC 1473-0024-1	13 53 43.79	+20 10 02.26	11.64	0.442
GSC 1473-0018	13 53 51.60	+20 08 18.60	12.29	0.569

Note: a. V-mag and (B–V) for comparison stars derived from MPOSC3 which is a hybrid catalog that includes a large subset of the Carlsberg Meridian Catalog *(CMC-14) as well as from the* Sloan Digital Sky Survey *(SDSS). Stars with BVI$_c$ magnitudes derived from 2MASS J–K magnitudes have an internal consistency of ± 0.05 mag. for V, ± 0.08 mag. for B, ± 0.03 mag. for I$_c$, and ± 0.05 mag. for B–V (Warner 2007).*

Table 2. Calculated differences (ETD)$_1$ following linear least squares fit of observed times-of-minimum for GW Boo and cycle number between 07 June 1999 and 28 June 2017.

HJD = 2400000+	Cycle No.	ETD$_1^a$	Reference
51336.7680	–2189	–0.01631	NSVS (Wozniak *et al.* 2004)[b]
51620.8784	–1654.5	–0.01639	NSVS (Wozniak *et al.* 2004)[b]
52750.6864	471	–0.00601	ASAS (Pojmański *et al.* 2005)[b]
52751.7460	473	–0.00950	Otero 2004
52755.4724	480	–0.00388	SAVS (Maciejewski *et al.* 2003)[c]
52767.4280	502.5	–0.00804	SAVS (Maciejewski *et al.* 2003)[c]
52788.4237	542	–0.00835	Maciejewski *et al.* 2003
53462.6857	1810.5	–0.01044	ASAS (Pojmański *et al.* 2005)[b]
54555.8090	3867	–0.00819	Diethelm 2010
55294.3962	5256.5	–0.00194	Hübscher and Monninger 2011
55310.4066	5286.5	0.06213	Hübscher and Monninger[d] 2011
55310.5829	5287	–0.02734	Hübscher and Monninger[d] 2011
55352.8640	5366.5	–0.00402	Diethelm 2010
55631.9288	5891.5	–0.00003	Diethelm 2011
55687.4748	5996	–0.00045	Hoňková K. *et al.* 2013
55698.3703	6016.5	–0.00158	Nagai 2012
55702.3597	6024	0.00123	Nagai 2012
55711.3947	6041	–0.00002	Nagai 2012
55715.6468	6049	–0.00032	This study
55719.6324	6056.5	–0.00124	This study
55720.6960	6058.5	–0.00079	This study
55749.6671	6113	0.00115	This study
56001.8846	6587.5	0.00087	Diethelm 2012
56056.3663	6690	–0.00078	Hoňková K. *et al.* 2013
56074.7055	6724.5	0.00018	Diethelm 2012
56418.3519	7371	0.00313	Hübscher 2013
56764.3830	8022	–0.00118	Hübscher and Lehmann 2015
57119.4516	8690	–0.00424	Hübscher 2017
57128.4893	8707	–0.00279	Hübscher 2017
57489.4087	9386	–0.00204	Hübscher 2017
57516.5166	9437	–0.00290	Hübscher 2017
57859.0953	10081.5	–0.00457	Nagai 2018
57914.6423	10186	–0.00392	This study
57919.6904	10195.5	–0.00552	This study
57931.6506	10218	–0.00508	This study
57932.7142	10220	–0.00456	This study

Notes: a. (ETD)$_1$ = Eclipse Time Difference between observed time-of-minimum and that calculated using the reference ephemeris (Equation 1). b. Interpolated from superimposition of NSVS, ASAS, and UO2017 lightcurves (see Figure 3). c. Times-of-minimum determined from BD+20°2890 lightcurves in SAVS database. d. Outliers not included in analysis.

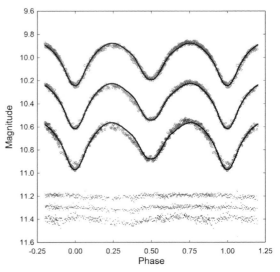

Figure 1. Folded CCD light curves for GW Boo produced from photometric data obtained between 03 June 2011 and 07 July 2011. The top (I$_c$), middle (V), and bottom curve (B) shown above were reduced to MPOSC3-based catalog magnitudes using MPO CANOPUS (Minor Planet Observer 2015). In this case, the Roche model assumed an A-type overcontact binary with no spots; residuals from the model fits are offset at the bottom of the plot to keep the values on scale.

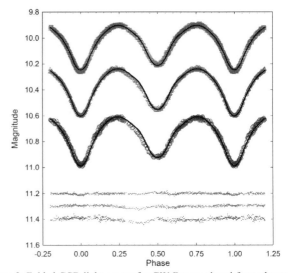

Figure 2. Folded CCD light curves for GW Boo produced from photometric data obtained between 06 June 2017 and 28 June 2017. The top (I$_c$), middle (V), and bottom curve (B) shown above were reduced to MPOSC3-based catalog magnitudes using MPO CANOPUS. In this case, the Roche model assumed an A-type overcontact binary with no spots; residuals from the model fits are offset at the bottom of the plot to keep the values on scale.

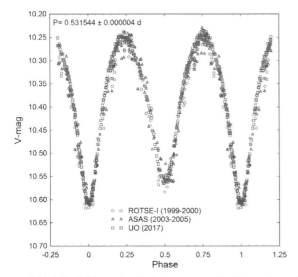

Figure 3. Folded (P = 0.531544 d) CCD light curves for GW Boo produced from sparsely sampled photometric data acquired during the ROTSE-I (1999–2000) and ASAS (2003–2005) surveys along with data generated at UO between 06 June 2017 and 28 June 2017.

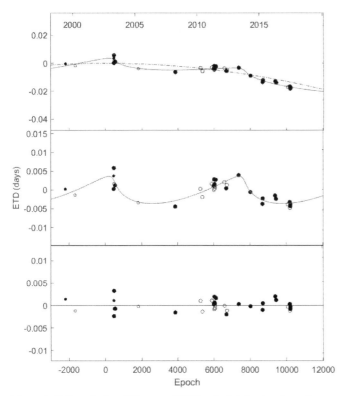

Figure 4. Eclipse timing differences (ETD) calculated using the reference ephemeris (Equation 2) cited by Kreiner (2004). LiTE analysis (top panel—solid line) for a putative third body where an elliptical path (e = 0.764) is predicted with an orbital period of 10.48 ± 0.01 y. The downwardly directed quadratic fit to the data is shown with a dashed line. Solid circles are from primary minima whereas open circles represent secondary minima. The center panel shows the LiTE fit after subtraction of the quadratic component. Residuals remaining from the model fit are plotted in the bottom panel.

An updated linear ephemeris (Equation 2) based on near-term (2013–2017) eclipse timing data was determined as follows:

$$\text{Min. I (HJD)} = 2457932.7138 \, (39) + 0.5315423 \, (4)\text{E}. \quad (2)$$

Secular variations in orbital period can sometimes be uncovered by plotting the difference between the observed eclipse times and those predicted by the reference epoch against cycle number (Figure 4). In this case the ETD residuals suggest there may be an underlying variability in the orbital period. This effect could potentially originate from magnetic cycles (Applegate 1992), the gravitational influence of a third body also known as the light-time effect (LiTE), or periodic mass transfer between either star. LiTE analysis was performed using the MATLAB (MathWorks©) code reported by Zasche *et al.* (2009) in which the associated parameters in the LiTE equation (Irwin 1959) were derived by simplex optimization. These include P_3 (orbital period of star 3 and the 1–2 pair about their common center of mass), orbital eccentricity *e,* argument of periastron *ω,* time of periastron passage T_0, and amplitude $A = a_{12} \sin i_3$ (where a_{12} = semimajor axis of the 1–2 pair's orbit about the center of mass of the three-star system, and i_3 = orbital inclination of the third body in a three-star system). For the sake of simplicity, a minimum mass for the putative third body was initially calculated after assuming a circular orbit (e = 0) which is co-planar (i_3 = 90°) with the binary pair. These results (LiTE-1) summarized in Table 3 suggest the presence of a stellar object with a mass approximating 0.21 $M_\odot$. According to tabulations by Harmanec (1988) and similar information on stellar mass by Pecaut and Mamajeck (2013) this third body is most likely an M-class star. A stellar object this small would only provide a slight excess in luminance ($L_3 < 0.07\%$); therefore no third light (l_3) contribution would be expected during Roche modeling of the light curves (section 3.4). The results (Table 3; Figure 4) with the lowest residual sum of squares (LiTE-2) predict a third body orbiting elliptically (e = 0.764 ± 0.197) every 10.48 ± 0.01 y. Similar to the case where e = 0, the fractional luminosity contributed by a gravitationally bound third body with minimum mass of 0.28 $M_\odot$ would still not reach significance during Roche modeling.

Alternatively, the sinusoidal variations in the orbital period of the binary pair may be due to magnetic activity cycles attributed to Applegate (1992). However, according to an empirical relationship (Equation 3) between the length of orbital period modulation and angular velocity ($\omega = 2\pi / P_{orb}$):

$$\log P_{mod}[y] = 0.018 - 0.36 \log (2\pi / P_{orb} \, [s] \quad (3)$$

(Lanza and Rodonò 1999) any period modulation resulting from a change in the gravitational quadrupole moment would probably be closer to 25 years for GW Boo, not the much shorter period ($P_3 < 10.5$ y) estimated from LiTE analyses.

Another revelation from the LiTE analysis is that the sign of the quadratic coefficient (c_2) is negative thereby indicating that the period is slowly decreasing with time. Based upon the results for LiTE-2 (Table 3), this translates into a orbital period decrease (dP/dt) approaching 1.8×10^{-7} d / y or 0.01556 s / y.

It is worth noting that eclipse timing data for GW Boo are only available for the past 18 years. This is not long enough to complete two cycles assuming that the proposed sinusoidal variability ($P_3 \sim 10$ y) is correct. As a result, careful examination of the data summarized in Table 3 reveals significant error in the LiTE-1 parameter estimates which became notably better for an elliptical orbit (LiTE-2). Another decade of eclipse timings will probably be needed to solidify a LiTE solution for this system.

3.2. Effective temperature estimation

Interstellar extinction (A_V) was estimated according to the model described by Amôres and Lépine (2005). In this case the value for A_V (0.085) corresponds to a target positioned at Galactic coordinates $l = 9.9936°$ and $b = +74.2456°$ which is located within 400 pc as determined from parallax (Gaia DR2: Brown *et al.* 2018). Color index (B–V) data collected at UO and those acquired from an ensemble of nine other sources (Table 4) were corrected using the estimated reddening value ($A_V / 3.1 =$ E(B–V) = 0.027 ± 0.001). The median intrinsic value (($B–V)_0 =$ 0.323 ± 0.012) which was adopted for Roche modeling indicates a primary star with an effective temperature (7080 K) that ranges in spectral type between F0V and F1V. This result is in good agreement with the Gaia DR2 release of stellar parameters (Andrae *et al.* 2018) in which the nominal T_{eff} for GW Boo is reported to be 6977 K. In contrast, an earlier study (Maciejewski *et al.* 2003) defines this system as a hotter ($T_{eff} = 7650$ K) and much larger spectral class A9III star based on an optical spectrum. According to the new results described herein, this classification is believed to be in error. It should be noted that luminosity classification of mid- to late A-type stars is especially difficult when trying to distinguish between dwarfs and giants (Gray and Corbally 2009). Furthermore, the mass of an A9 giant would approach 5 $M_\odot$ with a solar luminosity in excess of 26 $L_\odot$. As will be shown in the next section, the A9III assignment is rejected based on the mass, size, and luminosity obtained in this study and supported by other data included ($R_\odot$ and $L_\odot$) in the Gaia DR2 release of stellar parameters (Andrae *et al.* 2018).

3.3. Roche modeling approach

Roche modeling of LC data from GW Boo was primarily accomplished using the programs PHOEBE 0.31a (Prša and Zwitter 2005) and WDWINT 56a (Nelson 2009), both of which feature a user-friendly interface to the Wilson-Devinney WD2003 code (Wilson and Devinney 1971; Wilson 1990). WDWINT 56a makes use of Kurucz's atmosphere models (Kurucz 1993) which are integrated over $UBVR_cI_c$ optical passbands. In both cases, the selected model was Mode 3 for an overcontact binary. Bolometric albedo ($A_{1,2} = 0.5$) and gravity darkening coefficients ($g_{1,2} =$ 0.32) for cooler stars (7500 K) with convective envelopes were respectively assigned according to Ruciński (1969) and Lucy (1967). Since T_{eff} for the primary (7080 K) approaches the transition temperature where stars are in radiative equilibrium, modeling with $A_{1,2}$ and $g_{1,2}$ fixed at 1 was also explored. Logarithmic limb darkening coefficients (x_1, x_2, y_1, y_2) were interpolated (Van Hamme 1993) following any change in the effective temperature (T_{eff2}) of the secondary star during model fit optimization. All but the temperature of the more massive star (T_{eff1}), $A_{1,2}$ and $g_{1,2}$ were allowed to vary during DC iterations. In general, the best fits for T_{eff2}, i, q and Roche potentials ($\Omega_1 = \Omega_2$) were collectively refined (method of multiple subsets) by DC using the multicolor LC data. In general LCs from 2011 (Figures 1 and 5) and 2017 (Figures 2 and 6) exhibit significant asymmetry that is most obvious in the B-passband. This suggests the presence of spots (Yakut and Eggleton 2005) which were added during Roche modeling to address distorted/ asymmetric regions in the LCs.

3.4 Roche modeling results

GW Boo would appear to be an A-type overcontact system in which the primary star (m_1) is not only the more massive but also the hottest. The deepest minimum (Min I) occurs when the primary star is eclipsed by its smaller binary partner. These results are consistent with the general observation (Csizmadia and Klagyivik 2004; Skelton and Smits 2009) that A-type overcontact binaries have a mass ratio $m_2 / m_1 < 0.3$, are hotter than the Sun with spectral types ranging from A to F, and orbit the center-of-mass with periods varying between 0.4 to 0.8 d. Although a total eclipse is very nearly observed (Figures 7 and 8) there is some risk at attempting to determine a photometric mass ratio (q_{ptm}) by Roche modeling with the WD code alone (Terrell and Wilson 2005). They point out that even when the eclipses are "very slightly partial, the accuracy of a q_{ptm}

Table 3. Putative third-body solution to the light-time effect (LiTE) observed as sinusoidal-like changes in GW Boo eclipse timings.

Parameter	Units	LiTE-1	LiTE-2
HJD$_0$	—	2452500.3236 (9)	24552500.3242 (9)
P$_3$	[y]	9.84 (35)	10.48 (1)
A (semi-amplitude)	[d]	0.0034 (14)	0.0036 (4)
ω	°	—	146 (17)
e$_3$	—	0	0.764 (197)
a$_{12}$ sin i	[AU]	0.597 (250)	0.812 (86)
f(M$_3$)(mass function)	M$_\odot$	0.0022 (22)	0.0049 (1)
M$_3$ (i = 90°)	M$_\odot$	0.211 (161)	0.282 (1)
M$_3$ (i = 60°)	M$_\odot$	0.247 (72)	0.330 (2)
M$_3$ (i = 30°)	M$_\odot$	0.455 (144)	0.621 (3)
c$_2$ (quadratic coeff.)	~10^{-10}	−0.877 (1)	−1.31 (1)
dP/dt	10^{-7} d/yr	−1.205 (1)	−1.8 (1)
Sum of squared residuals	—	0.000531	0.000491

Table 4. Estimation of effective temperature (Teff1) of GW Boo based upon dereddened (B-V) data from six surveys, two published reports and the present study.

	USNO-B1.0	All Sky Combined	2MASS	APASS	Terrell et al. 2005	Tycho	UCAC4	Oja (1985)	Present Study
(B–V)$_0$	0.323	0.391	0.273	0.321	0.311	0.399	0.322	0.333	0.348
T$_{eff}$[a] (K)	7077	6715	7328	6622	7130	6685	7083	7036	6955
Spectral Class[a]	F0V-F1V	F3V-F4V	F0V-F1V	F4V-F5V	F0V-F1V	F3V-F4V	F0V-F1V	F0V-F1V	F1V-F2V

Note: a. T$_{eff1}$ interpolated and spectral class range estimated from Pecaut and Mamajek (2013). Median value, (B–V)$_0$ = 0.323 ± 0.012, corresponds to an F0V-F1V primary star (T$_{eff1}$ = 7080 ± 263 K).

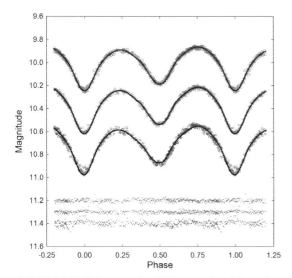

Figure 5. Folded CCD light curves for GW Boo produced from photometric data obtained between 03 June 2011 and 07 July2011. The top (I$_c$), middle (V), and bottom curve (B) shown above were reduced to MPOSC3-based catalog magnitudes using MPO CANOPUS. In this case, the Roche model assumed an A-type overcontact binary with a cool spot on the secondary star; residuals from the model fits are offset at the bottom of the plot to keep the values on scale.

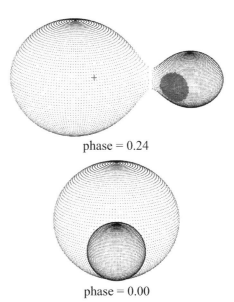

Figure 7. Spatial model of GW Boo from the 2011 LC (V-mag) illustrating the transit at Min I ($\varphi = 0$) and cool spot location ($\varphi = 0.24$) on the secondary star.

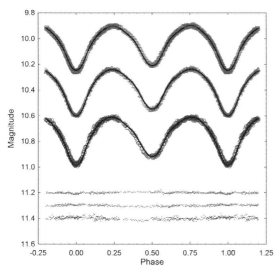

Figure 6. Folded CCD light curves for GW Boo produced from photometric data obtained between 08 June 2017 and 26 June 2017. The top (I$_c$), middle (V), and bottom curve (B) shown above were reduced to MPOSC3-based catalog magnitudes using MPO CANOPUS. In this case, the Roche model assumed an A-type overcontact binary with a single cool spot each on the primary and secondary stars; residuals from the model fits are offset at the bottom of the plot to keep the values on scale.

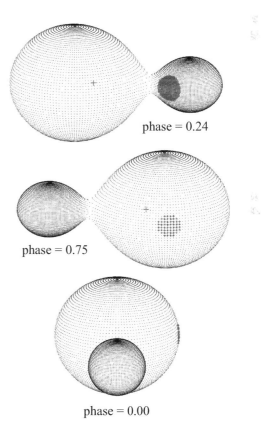

Figure 8. Spatial model of GW Boo from the 2017 LC (V-mag) showing the transit at Min I ($\varphi = 0$) and cool spot locations on the primary ($\varphi = 0.75$) and secondary stars ($\varphi = 0.24$).

drops dramatically." Given this proviso, modeling data collected with different equipment six years apart resulted in a mean best fit for the mass ratio (m_2/m_1) where q = 0.196 ± 0.007. Despite this good agreement, without the luxury of radial velocity data it is not possible to unequivocally determine the mass ratio for GW Boo or accurately establish the total mass.

Modeling under the assumption that GW Boo possesses a radiative envelope ($A_{1,2}$ and $g_{1,2}$ = 1) did not yield an improved fit compared to a system in convective equilibrium where $A_{1,2}$ and $g_{1,2}$ are respectively assigned values of 0.5 and 0.32. Only the latter LC parameters and geometric elements determined for each of these model fits (2011 and 2017) are summarized in Table 5. It should be noted that the listed errors only reflect the model fit to the observations which assumed exact values for all fixed parameters. The results are improbably low considering the estimated uncertainty (± 263 K) associated with the adopted T_{eff1} (Table 4) along with basic assumptions about $A_{1,2}$ and $g_{1,2}$ and the influence of putative spots added to the Roche model.

The fill-out parameter (*f*) which corresponds to a volume percent of the outer surface shared between each star was calculated according to Equation 4 (Kallrath and Milone 1999; Bradstreet 2005) where:

$$f = (\Omega_{inner} - \Omega_{1,2}) / (\Omega_{inner} - \Omega_{outer}). \quad (4)$$

Ω_{outer} is the outer critical Roche equipotential, Ω_{inner} is the value for the inner critical Roche equipotential and $\Omega = \Omega_{1,2}$ denotes the common envelope surface potential for the binary system. In this case the constituent stars are considered overcontact since 0 < *f* < 1. Spatial models rendered with BINARY BAKER3 (Bradstreet and Steelman 2004; using the physical and geometric elements from the best fit spotted Roche models are shown in Figure 7 (2011) and Figure 8 (2017).

3.5. Absolute parameters

Preliminary absolute parameters (Table 6) were derived for each star in this system using results from the best fit simulations (spotted model) of the 2011 and 2017 LCs. In the absence of RV data, total mass can not be unequivocally calculated; however, stellar mass and radii estimates from main sequence stars have been published over a wide range of spectral types. This includes a value (M_1 = 1.56 ± 0.07 $M_\odot$) interpolated from Harmanec (1988) and another (M_1 = 1.55 ± 0.04 $M_\odot$) from Pecaut and Mamajek (2013). Additionally, three different empirical period-mass relationships for W UMa-binaries have been published by Qian (2003) and later by Gazeas and Stępień (2008) and Gazeas (2009). According to Qian (2003) the mass of the primary star (M_1) can be determined from Equation 5:

$$\log M_1 = 0.761(150) \log P + 1.82(28), \quad (5)$$

where P is the orbital period in days and leads to M_1 = 1.73 ± 0.21 $M_\odot$ for the primary. The mass-period relationship (Equation 6) derived by Gazeas and Stępień (2008):

$$\log M_1 = 0.755(59) \log P + 0.416(24). \quad (6)$$

corresponds to a W UMa system where M_1 = 1.62 ± 0.11 $M_\odot$. Gazeas

(2009) reported another empirical relationship (Equation 7) for the more massive (M_1) star of a contact binary such that:

$$\log M_1 = 0.725(59) \log P - 0.076(32) \log q + 0.365(32). \quad (7)$$

In this case the mass for the primary star was estimated to be 1.66 ± 0.16 $M_\odot$. A final relationship reported by Torres *et al.* (2010) for main sequence stars above 0.6 $M_\odot$ predicts a mass of 1.55 $M_\odot$ for the primary constituent. The median of these six values (M_1 = 1.59 ± 0.04 $M_\odot$) was used for subsequent determinations of M_2, semi-major axis *a*, volume-radius r_L, bolometric magnitude M_{bol}, and ultimately distance *d* (pc) to GW Boo. The secondary mass = 0.31 ± 0.01 $M_\odot$ and total mass (1.90 ± 0.04 $M_\odot$) of the system were subsequently determined using the mean photometric mass ratio (0.196 ± 0.007). By comparison, a stand-alone main sequence star with a mass similar to the secondary (early M-type) would likely be much smaller ($R_\odot \sim 0.4$), cooler ($T_{eff} \sim 3600$), and far less luminous ($L_\odot \sim 0.03$). The mean semi-major axis, a($R_\odot$) = 3.42 ± 0.03, was calculated from Newton's version (Equation 8) of Kepler's third law where:

$$a^3 = (G \times P^2 (M_1 + M_2)) / (4\pi^2). \quad (8)$$

The effective radii of each Roche lobe (r_L) can be calculated to over the entire range of mass ratios (0 < q < ∞) according to an expression (Equation 9) derived by Eggleton (1983):

$$r_L = (0.49q^{2/3}) / (0.6q^{2/3} + ln(1 + q^{1/3})). \quad (9)$$

from which values for r_1 (0.5212 ± 0.0001) and r_2 (0.2514 ± 0.0001) were determined for the primary and secondary stars, respectively. Since the semi-major axis and the volume radii are known, the solar radii for both binary constituents can be calculated where R_1 = a · r_1 = (1.79 ± 0.01 $R_\odot$) and R_2 = a · r_2 = (0.85 ± 0.01 $R_\odot$).

Luminosity in solar units ($L_\odot$) for the primary (L_1) and secondary stars (L_2) were calculated from the well-known relationship (Equation 10) where:

$$L_{1,2} = (R_{1,2} / R_\odot)^2 (T_{1,2} / T_\odot)^4. \quad (10)$$

Assuming that T_{eff1} = 7080 K, mean T_{eff2} = 6689 K and $T_\odot$ = 5772 K, then the solar luminosities for the primary and secondary are L_1 = 7.27 ± 0.11 and L_2 = 1.37 ± 0.02, respectively. According to the Gaia DR2 release of stellar parameters (Andrae *et al.* 2018), T_{eff} (6977 K) is slightly cooler than the adopted T_{eff1} (7080 K) while the size ($R_\odot$ = 1.92) and luminosity ($L_\odot$ = 7.90) of the primary star in GW Boo are slightly greater than the values estimated by this study. By any measure these results are far removed from those expected from the A9III classification proposed by Maciejewski *et al.* (2003).

3.6. Distance estimates to GW Boo

The bolometric magnitudes ($M_{bol1,2}$) for the primary and secondary were determined according to Equation 11 such that:

$$M_{bol1,2} = 4.75 - 5 \log(R_{1,2} / R_\odot) - 10 \log(T_{1,2} / T_\odot). \quad (11)$$

Table 5. Synthetic light curve parameters evaluated by Roche modeling and the geometric elements derived for GW Boo, an A-type W UMa variable.

Parameter	2011 No spot	2011 Spotted	2017 No spot	2017 Spotted
T_{eff1} (K)[b]	7080	7080	7080	7080
T_{eff2} (K)	6662 ± 8	6606 ± 6	6766 ± 4	6771 ± 3
q (m_2 / m_1)	0.196 ± 0.001	0.206 ± 0.001	0.190 ± 0.001	0.192 ± 0.001
A[b]	0.5	0.5	0.5	0.5
g[b]	0.32	0.32	0.32	0.32
$\Omega_1 = \Omega_2$	2.190 ± 0.002	2.210 ± 0.002	2.188 ± 0.001	2.184 ± 0.001
$i°$	75.03 ± 0.19	75.03 ± 0.14	73.83 ± 0.09	74.67 ± 0.09
$A_S = T_S / T_\star$[c]	—	—	—	0.86 ± 0.01
Θ_S (spot co—latitude)[c]	—	—	—	90 ± 1.5
φ_S (spot longitude)[c]	—	—	—	95 ± 2
r_S (angular radius)[c]	—	—	—	11 ± 0.1
$A_S = T_S / T_\star$[d]	—	0.75 ± 0.01	—	0.80 ± 0.01
Θ_S (spot co—latitude)[d]	—	95.5 ± 2.1	—	86.8 ± 1.6
φ_S (spot longitude)[d]	—	59.4 ± 2.2	—	49.1 ± 2.3
r_S (angular radius)[d]	—	30 ± 0.6	—	25 ± 0.4
$L_1 / (L_1 + L_2)_B$[e]	0.8534 ± 0.0003	0.8529 ± 0.0003	0.8472 ± 0.0001	0.8444 ± 0.0001
$L_1 / (L_1 + L_2)_V$	0.8445 ± 0.0001	0.8425 ± 0.0001	0.8407 ± 0.0001	0.8378 ± 0.0001
$L_1 / (L_1 + L_2)_{Ic}$	0.8334 ± 0.0001	0.8297 ± 0.0001	0.8328 ± 0.0001	0.8296 ± 0.0001
r_1 (pole)	0.4964 ± 0.0001	0.4937 ± 0.0001	0.4956 ± 0.0001	0.4971 ± 0.0001
r_1 (side)	0.5431 ± 0.0002	0.5396 ± 0.0002	0.5417 ± 0.0001	0.5439 ± 0.0001
r_1 (back)	0.5678 ± 0.0003	0.5651 ± 0.0002	0.5653 ± 0.0001	0.5684 ± 0.0001
r_2 (pole)	0.2399 ± 0.0008	0.2447 ± 0.0005	0.2347 ± 0.0003	0.2383 ± 0.0003
r_2 (side)	0.2507 ± 0.0009	0.2559 ± 0.0007	0.2448 ± 0.0004	0.2489 ± 0.0003
r_2 (back)	0.2915 ± 0.0020	0.2980 ± 0.0015	0.2820 ± 0.0008	0.2891 ± 0.0007
Fill—out factor (%)	25.6	28.4	16.2	24.2
RMS (B)[f]	0.02414	0.02012	0.01295	0.01102
RMS (V)[f]	0.01586	0.01175	0.00853	0.00700
RMS (I$_c$)[f]	0.01500	0.01200	0.00647	0.00630

Notes: a. All error estimates for T_{eff2}, q, $\Omega_{1,2}$, A_S, Θ_S, φ_S, r_S, $r_{1,2}$, and L_1 from WDWINT 56a (Nelson 2009).
b. Fixed during DC.
c. Primary spot temperature, location and size parameters in degrees.
d. Secondary spot temperature, location and size parameters in degrees.
e. L_1 and L_2 refer to scaled luminosities of the primary and secondary stars, respectively.
f. Monochromatic root mean square deviation of model fit from observed values (mag).

Table 6. Preliminary absolute parameters (± SD) for GW Boo using the mean photometric mass ratio ($q_{ptm} = m_2 / m_1$) from the Roche model fits of LC data (2011 and 2017) and estimated mass for an F0V-F1V primary star.

Parameter	Primary	Secondary
Mass ($M_\odot$)	1.59 ± 0.04	0.31 ± 0.01
Radius ($R_\odot$)	1.79 ± 0.01	0.85 ± 0.01
a ($R_\odot$)	3.42 ± 0.03	—
Luminosity ($L_\odot$)	7.27 ± 0.11	1.37 ± 0.02
M_{bol}	2.60 ± 0.02	4.41 ± 0.02
Log (g)	4.13 ± 0.01	4.06 ± 0.01

This led to values where $M_{bol1} = 2.61 \pm 0.01$ and $M_{bol2} = 4.42 \pm 0.01$. Combining the bolometric magnitudes resulted in an absolute magnitude ($M_V = 2.43 \pm 0.01$) after adjusting with the bolometric correction (BC = –0.010) interpolated from Pecaut and Mamajek (2013). Substituting into the distance modulus (Equation 12):

$$d(pc) = 10^{((m - Mv - Av + 5)/5)}, \qquad (12)$$

where m = V_{max} (10.23 ± 0.01) and $A_V = 0.085$ leads to an estimated distance of 349 ± 2 pc to GW Boo. This value is about 12% lower than the distance (398 ± 9 pc) calculated directly from the second release (DR2) of parallax data from the Gaia mission (Lindegren et al. 2016; Brown et al. 2018). Considering that LC magnitudes were not determined using absolute photometry and perhaps more importantly the physical size/luminosity were not derived from a spectroscopic (RV) mass ratio, this discrepancy is not unreasonable.

4. Conclusions

Eight new times-of-minimum were observed based on CCD data collected with B, V, and I$_c$ filters. These along with other published values led to an updated linear ephemeris for GW Boo. Potential changes in orbital periodicity were assessed using eclipse timings which only cover a 18-year time span. Nevertheless, an underlying sinusoidal variability ($P_3 \sim 10.5$ y) in the orbital period was uncovered which suggests the possibility of a third gravitationally bound but much smaller stellar object. The intrinsic color, $(B–V)_0$, determined from this study and eight other sources indicates that the effective temperature for the primary is ~ 7080 K which corresponds to a spectral class ranging from F0V to F1V. This A-type overcontact system very nearly experiences a total eclipse. Therefore the photometric mass ratio determined by Roche modeling (q = 0.196 ± 0.007) may suffer in accuracy based on precautions noted by Terrell and Wilson (2005). Radial velocity findings will be required to unequivocally determine a spectroscopically derived mass ratio and total mass for the system. GW Boo is bright enough ($V_{mag} \sim 10.2$) to be considered a viable candidate for further spectroscopic study to unequivocally classify this system and generate RV data to refine (or refute) the absolute parameter values presented herein.

5. Acknowledgements

This research has made use of the SIMBAD database operated at Centre de Données astronomiques de Strasbourg, France. Time-of-minima data tabulated in the Variable Star Section of the Czech Astronomical Society (B.R.N.O.) website proved invaluable to the assessment of potential period changes experienced by this variable star. In addition, the Northern Sky Variability Survey hosted by the Los Alamos National Laboratory, the International Variable Star Index maintained by the AAVSO, the Semi-Automatic Variability Search at the Nicolaus Copernicus University, and the ASAS Catalogue of Variable Stars were mined for photometric data. The diligence and dedication shown by all associated with these organizations is very much appreciated. This work also presents results from the European Space Agency (ESA) space mission Gaia. Gaia data are being processed by the Gaia Data Processing and Analysis Consortium (DPAC). Funding for the DPAC is provided by national institutions, in particular the institutions participating in the Gaia MultiLateral Agreement (MLA). The Gaia mission website is https://www.cosmos.esa.int/gaia. The Gaia archive website is https://archives.esac.esa.int/gaia. Furthermore, the author greatly appreciates the review, corrections and suggested changes made by an anonymous referee.

References

Akerlof, C., et al. 2000, *Astron. J.*, **119**, 1901.

Amôres, E. B., and Lépine, J. R. D. 2005, *Astron. J.*, **130**, 650.

Andrae, R., et al. 2018, *Astron. Astrophys.*, **616A**, 8.

Applegate, J. H. 1992, *Astrophys. J.*, **385**, 621.

Berry, R., and Burnell, J. 2005, *The Handbook of Astronomical Image Processing*, 2nd ed., Willmann-Bell, Richmond, VA.

Bradstreet, D. H. 2005, in *The Society for Astronomical Sciences 24th Annual Symposium on Telescope Science*, Society for Astronomical Sciences, Rancho Cucamonga, CA, 23.

Bradstreet, D. H., and Steelman, D. P. 2004, BINARY MAKER 3, Contact Software (http://www.binarymaker.com).

Brown, A. G. A., et al. 2018, *Astron. Astrophys.*, **616A**, 1.

Csizmadia, Sz., and Klagyivik, P. 2004, *Astron. Astrophys.*, **426**, 1001.

Diethelm, R. 2010, *Inf. Bull. Var. Stars*, No. 5945, 1.

Diethelm, R. 2011, *Inf. Bull. Var. Stars*, No. 5992, 1.

Diethelm, R. 2012, *Inf. Bull. Var. Stars*, No. 6029, 1.

Eggleton, P. P. 1983, *Astrophys. J.*, **268**, 368.

Gazeas, K. D. 2009, *Commun. Asteroseismology*, **159**, 129.

Gazeas, K., and Stępień, K. 2008, *Mon. Not. Roy. Astron. Soc.*, **390**, 1577.

Gettel, S. J., Geske, M. T., and McKay, T. A. 2006, *Astron. J.*, **131**, 621.

Gray, R. O., and Corbally, C. J. 2009, *Stellar Spectral Classification*, Princeton Univ. Press, Princeton, NJ.

Harmanec, P. 1988, *Bull. Astron. Inst. Czechoslovakia*, **39**, 329.

Harris, A. W., et al. 1989, *Icarus*, **77**, 171.

Hoňková K., et al. 2013, *Open Eur. J. Var. Stars*, **160**, 1.

Hübscher, J. 2013, *Inf. Bull. Var. Stars*, No. 6084, 1.

Hübscher, J. 2017, *Inf. Bull. Var. Stars*, No. 6196, 1.

Hübscher, J., and Lehmann, P. B. 2015, *Inf. Bull. Var. Stars*, No. 6149, 1.

Hübscher, J., and Monninger, G. 2011, *Inf. Bull. Var. Stars*, No. 5959, 1.

Irwin, J. B. 1959, *Astron. J.*, **64**, 149.

Kallrath, J., and Milone, E. F. 1999, *Eclipsing Binary Stars: Modeling and Analysis*, Springer, New York.

Kreiner, J. M. 2004, *Acta Astron.*, **54**, 207.

Kurucz, R. L. 1993, *Light Curve Modeling of Eclipsing Binary Stars*, ed. E. F. Milone, Springer-Verlag, New York, 93.

Kwee, K. K., and van Woerden, H. 1956, *Bull. Astron. Inst. Netherlands*, 12, 327.

Lanza, A. F., and Rodonò, M. 1999, *Astron. Astrophys.*, **349**, 887.

Lindegren, L., et al. 2016, *Astron. Astrophys.*, **595A**, 4.

Lucy, L. B. 1967, *Z. Astrophys.*, **65**, 89.

Maciejewski, G., Czart, K., Niedzielski, A., and Karska, A. 2003, *Inf. Bull. Var. Stars*, No. 5431, 1.

Nagai. K. 2012, *Bull. Var. Star Obs. League Japan*, No. 53, 1.

Nagai. K. 2018, *Bull. Var. Star Obs. League Japan*, No. 64, 1.

Nelson, R. H. 2009, WDWINT version 56a astronomy software (https://www.variablestarssouth.org/bob-nelson).

Otero, S. A. 2004, *Inf. Bull. Var. Stars*, No. 5532, 1.

Paunzen, E., and Vanmunster, T. 2016, *Astron. Nachr.*, **337**, 239.

Pecaut, M. J., and Mamajek, E. E. 2013, *Astrophys. J. Suppl. Ser.*, **208**, 9.

Pojmański, G., Pilecki, B., and Szczygiel, D. 2005, *Acta Astron.*, **55**, 275.

Prša, A., and Zwitter, T. 2005, *Astrophys. J.*, **628**, 426.

Qian, S.-B. 2003, *Mon. Not. Roy. Astron. Soc.*, **342**, 1260.

Ruciński, S. M. 1969, *Acta Astron.*, **19**, 245.

Schwarzenberg-Czerny, A. 1996, *Astrophys. J., Lett.*, **460**, L107.

Skelton, P. L., and Smits, D. P. 2009, *S. Afr. J. Sci.*, **105**, 120.

Terrell, D., and Wilson, R. E. 2005, *Astrophys. Space Sci.*, **296**, 221.

Torres, G., Andersen, J., and Giménez, A. 2010, *Astron. Astrophys. Rev.*, 18, 67.

Van Hamme, W. 1993, *Astrophys. J.*, **106**, 2096.

Warner, B. 2007, *Minor Planet Bull.*, **34**, 113.

Wilson, R. E. 1990, *Astrophys. J.*, **356**, 613.

Wilson, R. E., and Devinney, E. J. 1971, *Astron. J.*, **143**, 1.

Wozniak, P. R., et al. 2004, *Astron. J.*, **127**, 2436.

Yakut, K., and Eggleton, P. P. 2005, *Astrophys. J.*, **629**, 1055.

Zasche, P., Liakos, A., Niarchos, P., Wolf, M., Manimanis, V., and Gazeas, K. 2009, *New Astron.*, **14**, 121.

Photometry of Fifteen New Variable Sources Discovered by IMSNG

Seoyeon Choi
Phillips Academy, 180 Main Sreet, Andover, MA 01810; schoi1@andover.edu

Myungshin Im
Center for the Exploration of the Universe (CEOU), Astronomy Program, Department of Physics and Astronomy, Seoul National University, Seoul 08826, Korea; mim@astro.snu.ac.kr

Changsu Choi
Center for the Exploration of the Universe (CEOU), Astronomy Program, Department of Physics and Astronomy, Seoul National University, Seoul 08826, Korea; changsu@astro.snu.ac.kr

Received September 25, 2018; revised October 18, 22, November 11 2018; accepted November 12, 2018

Abstract   We report the discovery of fifteen new variable objects from data taken in the course of a survey that monitors nearby galaxies to uncover the onset of supernovae. A light curve was generated for each variable star candidate and was evaluated for variability in brightness and periodicity. Three objects were determined as periodic variables through period analysis using VSTAR. Nearly periodic, short-duration dips are found for three objects, and these objects are likely to be eclipsing binaries. Variability of the remaining sources are rather random, and we were not able to conclude whether they are irregular variables or not from a lack of data.

1. Introduction

Intensive Monitoring Survey of Nearby Galaxies (IMSNG) is a daily monitoring program of 60 nearby galaxies to catch the early light curve of supernovae operated by the Center for the Exploration of the Origin of the Universe (CEOU), Seoul National University (Im *et al.* 2015a). While IMSNG's principle purpose has been observing supernovae, the high cadence of the survey has provided opportunities for the discovery of variable stars. In this paper, we present the photometry and period analysis of fifteen previously unknown variable stars using IMSNG data.

Observation information and data calibration techniques are discussed in section 2. A light curve was generated for each variable star candidate to be evaluated for variability in brightness and periodicity; the photometry and period analysis of the variable star candidates are presented and analyzed in section 3. We conclude in section 4.

2. Observation and data calibration

For this paper, we used *r*-band images taken with the Lee Sang Gak Telescope (LSGT hereafter; Im *et al.* 2015b for LSGT paper), a 0.43-m telescope at the Siding Springs Observatory, Australia. The images were taken with SNUCAM-II camera which provides a field of view of 15.7' × 15.7' and a pixel scale of 0.92" (Choi and Im 2017). Under normal circumstances, each field containing a target galaxy at the center was imaged once a day during the observation period. A total of three frames were taken at a given epoch and were later combined into a single, deeper image. Reduction of LSGT data was done by dark subtraction and flat-fielding.

In order to identify transients easily, we created a reference image to subtract from the LSGT data. For each field, eight to fifteen nights of LSGT data taken between the first two months

of observation were selected and combined into a master reference. Images with high zero-point and low full-width-half-maximum (FWHM) values were chosen. The reference image went through a process of convolution with a Gaussian profile and flux-scaling to match the seeing and the zero-point of each image.

The subtraction yields an image in which sources with constant brightness are erased out of the image. In comparison to the reference image, sources that had increased in brightness appear as white dots and ones that had decreased in brightness appear as black dots on the image. Sources that switched between white and black dots on the subtracted image were selected as variable star candidates. Subtracted images of variable star candidate USNO-B1.0 0685-0078225 are presented in Figure 1. The variable star candidates were cross-identified using NED (NASA/IPAC Extragalactic database) and IRSA (NASA/IPAC Infrared Science Archive). All the objects discussed in this paper are not identified as variables in the VSX (The International Variable Star Index).

Figure 1. Subtracted images of USNO-B1.0 0685-0078225. The right image is from Julian Date (JD) 2457685 and the left is from JD 2457696.

After this visual inspection, the standard deviation of the light curve is compared with that of the average photometric error of the object. When the light curve standard deviation is 2.56 times that of the average photometric error, we identify the object as variable. The factor of 2.56 nominally represents 99% confidence in the reality of the variability according to the C-test (Jang and Miller 1997; Romero *et al.* 1999; also see Kim *et al.* 2018).

Table 1. Variable sources identified from our data. Identifications, positional data, and classifications were taken from NED and IRSA.

Identification	R. A. (2000) h m s	Dec. (2000) ° ' "	Mag.[1]	Mag. error[2]	Variability Mag.[3]	Classification
STSSL2 J041928.18-545750.708	04 19 28.185	–54 57 50.73	16.407	0.019	0.717	IrS[4]
GALEXMSC J042034.71-550055.708	04 20 34.72	–55 00 55.7	15.259	0.016	0.271	IrS, UvS5
USNO-B1.0 0685-0078225	06 16 40.50	–21 25 01.23	15.425	0.020	0.300	IrS
2MASS J10253022-3952044	10 25 30.225	–39 52 04.48	17.344	0.033	0.484	IrS
2MASS J10260936-3947373	10 26 09.369	–39 47 37.32	17.138	0.031	1.512	IrS
GALEXASC J102502.22-294913.8	10 25 02.22	–39 49 13.8	13.195	0.012	0.601	IrS, UvS
GALEXASC J102509.56-394736.7	10 25 09.57	–39 47 36.7	14.951	0.014	0.559	IrS, UvS
SDSS J123801.86+115436.5	12 38 01.865	+11 54 36.54	15.177	0.018	0.285	IrS[6]
GALEXMSC J180742.51+174200.4	18 07 42.51	+17 42 00.4	16.319	0.030	0.510	UvS
GALEXASC J180759.71+173815.4	18 07 59.71	+17 38 15.4	16.684	0.034	0.321	IrS, UvS
2MASS J18080948+1736557	18 08 09.49	+17 36 55.77	16.82	0.037	0.410	IrS, UvS
2MASS J19422267-1019583	19 42 22.673	–10 19 58.33	16.542	0.034	0.433	IrS, UvS
GALEXASC J194247.24-102215.3	19 42 47.24	–10 22 15.4	14.981	0.028	0.338	IrS, UvS
SSTSL2 J194231.40-102150.5	19 42 31.409	–10 21 50.52	16.62	0.035	0.441	IrS
2MASS J18180055-5443534	18 18 00.559	–54 43 53.20	15.722	0.030	0.896	IrS

Notes: 1. median; 2. average; 3. max magnitude – min magnitude; 4. infrared source; 5. ultraviolet source; 6. star.

We selected three calibration stars, which also served as comparison stars for the variable star candidates, from the data release 8 (DR8) of the AAVSO Photometric All-Sky Survey (APASS; Henden *et al.* 2012) for the calculation of the photometric zero point for each epoch data. Stars with *r*-band magnitude error = 0, r < 14 mag, r > 17 mag, or a nearby light source were eliminated in the selection process. The stars with r < 14 mag are found to be saturated at center in the LSGT images, and the stars with r > 17 mag have large photometric errors in the APASS catalog. Values for *r*-band magnitude and *r*-band magnitude error were directly taken from APASS. The magnitudes of the objects were measured using a 3.0-inch diameter aperture with SEXTRACTOR (Bertin and Arnouts 1996) on images before the subtraction of the reference image. Period analysis on the variable star candidates was performed using VSTAR, developed by the AAVSO.

3. Results and discussion

Here, we describe the properties of each variable sources we identified. Table 1 describes the summary of the newly identified variable sources.

3.1. STSSL2 J041928.18-545750.7

This object was observed in the direction of NGC 1566. The cross-identifications of the object are listed in Table 2. It is classified as an infrared source in NED. Figure 2 is a light curve of the object, which was imaged over a nine-month span from Heliocentric Julian Date (HJD) 2457595 to 2457862. The brightness range over the observed period is ~0.7 magnitude. Figure 3 is a light curve of STSSL2 J041928.18-545750.7 and its comparison stars. The information about the comparison stars is listed in Table 3. While the comparison stars exhibit no variability in brightness, the variable star candidate shows strong variability. From our period analysis, the object appeared to have no signs of periodicity. However, it is likely that the result was influenced by a lack of data. With observations conducted only once a night, at its highest frequency, it is difficult for our data to reveal periods less than one day.

It cannot be concluded whether STSSL2 J041928.18-545750.7 is an irregular or periodic variable without further observations.

Table 2. Cross-identifications of STSSL2 J041928.18-545750.7. Positional data were taken from NED and IRSA.

Identification	R. A. (2000) h m s	Dec. (2000) ° ' "
STSSL2 J041928.18-545750.7	04 19 28.185	–54 57 50.73
2MASS J04192819-5457506	04 19 28.193	–54 57 50.69
USNO-B1.0 0350-0032627	04 19 28.15	–54 57 50.59

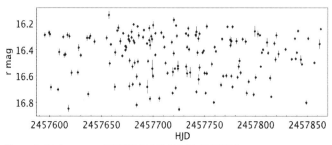

Figure 2. Light curve of STSSL2 J041928.18-545750.7.

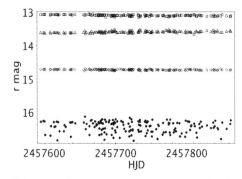

Figure 3. Light curve of STSSL2 J041928.18-545750.7 (solid circle) and its comparison stars (c1, open circle; c2, square; c3, triangle). The error bars are omitted.

Table 3. Comparison stars for variable star candidates in the field of NGC1566: STSSL2 J041928.18-545750.7 and GALEXMSC J042034.71-550055.7. Data were taken from APASS.

Name[1]	Record No.[2]	R. A. (2000) h m s	Dec. (2000) ° ' "	r	r error
Comparison 1	36757548	04 20 23.51	−55 00 25.41	14.717	0.031
Comparison 2	36757545	04 20 22.18	−55 02 42.55	13.104	0.027
Comparison 3	36757550	04 19 48.85	−54 59 58.35	13.623	0.026

Notes: 1. in this paper; 2. in APASS.

3.2. GALEXMSC J042034.71-550055.7

This object shares the same field as STSSL2 J041928.18-545750.7. It is classified as an infrared and ultraviolet source in NED. The cross-identifications of the object are listed in Table 4. Figure 4 is a light curve of the object, which was imaged from HJD 2457595 to 2457862. The brightness range over the observed period is ~0.2 magnitude. Figure 5 is a light curve of GALEXMSC J042034.71-550055.7 and its comparison stars. The information about the comparison stars is listed in Table 3. The object shows a stronger sign of variability in brightness than its comparison stars. A period analysis revealed three possible periods of variability: 2.701 days and its aliases, 1.585 days and 0.729 days. Figure 6 is the power spectrum and Figure 7 is the phase plot phased to a period of 2.701 days.

Table 4. Cross-identifications of GALEXMSC J042034.71-550055.7. Positional data were taken from NED and IRSA.

Identification	R. A. (2000) h m s	Dec. (2000) ° ' "
GALEXMSC J042034.71-550055.7	04 20 34.72	−55 00 55.7
2MASS J04203483-5500557	04 20 34.836	−55 00 55.76
SSTSL2 J042043.84-550055.7	04 20 34.846	−55 00 55.73
USNO-B1.0 0349-0032686	04 20 34.81	−55 00 55.63

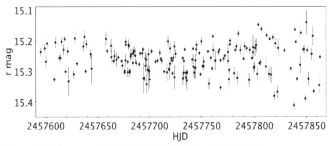

Figure 4. Light curve of GALEXMSC J042034.71-550055.7.

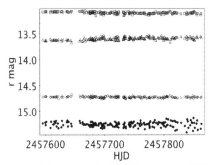

Figure 5. Light curve of GALEXMSC J042034.71-550055.7 (solid circle) and its comparison stars (c1, open circle; c2, square; c3, triangle). Error bars are omitted.

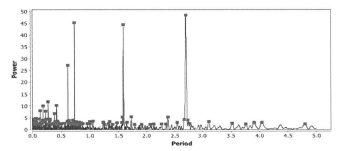

Figure 6. Power spectrum for GALEXMSC J042034.71-550055.7. Period is in days.

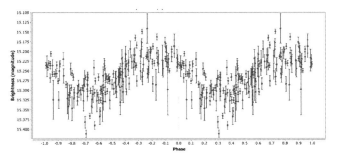

Figure 7. Phase plot of GALEXMSC J042034.71-550055.7 phased to a period of 2.701 days. Brightness (vertical axis) is in r magnitude.

3.3. USNO-B1.0 0685-0078225

This object was observed in the field of NGC 2207. It is classified as an infrared source in NED. The cross-identifications of the object are listed in Table 5. Figure 8 is a light curve of USNO-B1.0 0685-0078225 and its comparison stars, which were imaged from HJD 2457609 to 2457889. The information about the comparison stars is listed in Table 6. The brightness range over the observed period is ~0.25 magnitude. The object shows a strong evidence of variability in brightness and periodicity; a sinusoidal light curve of the variable star candidate could be observed. Period analysis on the object revealed a period of 86.337 days. Figure 9 is a phase diagram of USNO-B1 0685-0078225.

Table 5. Cross-identifications of USNO-B1.0 0685-0078225. Positional data were taken from NED and IRSA.

Identification	R. A. (2000) h m s	Dec. (2000) ° ' "
USNO-B1.0 0685-0078225	06 16 40.50	−21 25 01.23
2MASS J06164051-2125010	06 16 40.512	−21 25 01.08
SSTSL2 J061640.49-212501.0	06 16 40.494	−21 25 01.10

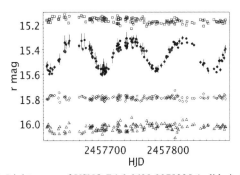

Figure 8. Light curve of USNO-B1.0 0685-0078225 (solid circle) and its comparison stars (c1, open circle; c2, square; c3, triangle).

Table 6. Comparison stars for USNO-B1.0 0685-0078225. Data were taken from APASS.

Name[1]	Record No.[2]	R. A. (2000) h m s	Dec. (2000) ° ' "	r	r error
Comparison 1	18952833	06 16 44.32	–21 27 7.19	15.819	0.041
Comparison 2	18953603	06 16 45.82	–21 17 40.48	15.201	0.019
Comparison 3	18953333	06 16 21.51	–21 26 40.87	15.97	0.039

Notes: 1. in this paper.; 2. in APASS.

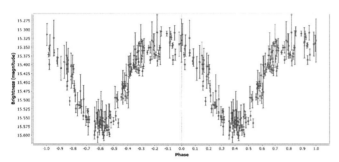

Figure 9. Phase plot of USNO-B1.0 0685-0078225 phased to a period of 86.337 days. Brightness (vertical axis) is in r magnitude.

3.4. 2MASS J10253022-3952044

This object was observed in the field of NGC 3244. It is classified as an infrared source in NED. The cross-identifications of the object are listed in Table 7. Figure 10 is a light curve of the object containing data points from HJD 2457690 to 2457942, which is about eight months. Figure 11 is a light curve of the object and its comparison stars. Table 8 includes information about the comparison stars. The brightness range over the observed period is ~0.5 magnitude. The variable star candidate exhibits an irregular variability in brightness and a period analysis was not able to identify a period. However, the object shows occasional dips in magnitude, and it is likely to be an eclipsing binary star system. Further observations should be conducted to reveal the complete behavior of the object.

Table 7. Cross-identifications of 2MASS J10253022-3952044. Positional data were taken from NED and IRSA.

Identification	R. A. (2000) h m s	Dec. (2000) ° ' "
2MASS J10253022-3952044	10 25 30.225	–39 52 04.48
USNO-B1.0 0501-0215795	10 25 30.24	–39 52 04.71

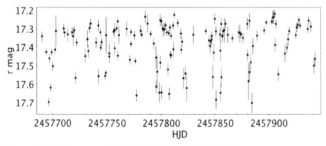

Figure 10. Light curve of 2MASS J10253022-3952044.

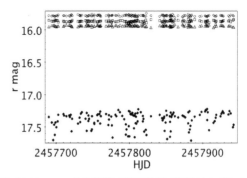

Figure 11. Light curve of 2MASS J10253022-3952044 (solid circle) and its comparison stars (c1, open circle; c2, square; c3, triangle). Error bars are omitted.

Table 8. Comparison stars for variable star candidates in the field of NGC 3244: 2MASS J10253022-3952044, 2MASS J10260936-3947373, GALEXASC J102502.22-294913.8, and GALEXASC J102509.56-394736.7. Data were taken from APASS.

Name[1]	Record No.[2]	R. A. (2000) h m s	Dec. (2000) ° ' "	r	r error
Comparison 1	42498874	10 25 13.69	–39 44 11.05	15.754	0.06
Comparison 2	42498886	10 25 10.77	–39 42 10.31	15.918	0.022
Comparison 3	42498875	10 25 11.82	–39 42 37.4	15.92	0.034

Notes: 1. in this paper; 2. in APASS.

3.5. 2MASS J10260936-3947373

This object was observed in the same field of 2MASS J10253022-3952044. It is classified as an infrared source in NED. The cross-identifications of the object are listed in Table 9. Figure 12 is a light curve of the object which was imaged from HJD 2457690 to 2457941. Figure 13 is a light curve of 2MASS J10260936-3947373 and its comparison stars. Table 8 includes information about the comparison stars. The brightness range over the observed period is ~1.0 magnitude. No periodicity of the object was identified in the period analysis, but just like 2MASS J10253022-3952044, the occasional, short-duration dips in brightness suggest that this object is probably an eclipsing binary. Further observations should be conducted of this object.

Table 9. Cross-identifications of 2MASS J10260936-3947373. Positional data were taken from NED and IRSA.

Identification	R. A. (2000) h m s	Dec. (2000) ° ' "
2MASS J10260936-3947373	10 26 09.369	–39 47 37.32
USNO-B1.0 0502-0215697	10 26 09.38	–39 47 37.60

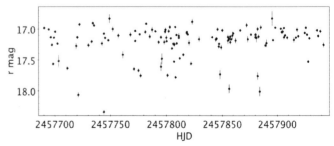

Figure 12. Light curve of 2MASS J10260936-3947373.

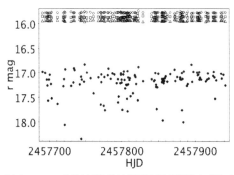

Figure 13. Light curve of 2MASS J10260936-3947373 (solid circle) and its comparison stars (c1, open circle; c2, square; c3, triangle). Error bars are omitted.

3.6. GALEXASC J102502.22-294913.8

This object was observed in the same field as the previous object. It is classified as an infrared and ultraviolet source in NED. The cross-identifications of the object are listed in Table 10. Figure 14 is a light curve of GALEXASC J102502.22-294913.8, which was imaged from HJD 2457690 to 2457942. Figure 15 is a light curve of the object and its comparison stars. Information about the comparison stars is listed in Table 8. The brightness range over the observed period is ~0.5 magnitude. The brightness of the object remains relatively constant except for when an unusual darkening occurs every ~70 days. This periodic dip in brightness suggests that this object is an eclipsing binary star system. However, when performing a period analysis using VSTAR, we failed to identify a periodicity. Further observations should be conducted of this object to confirm its pattern of variability in brightness.

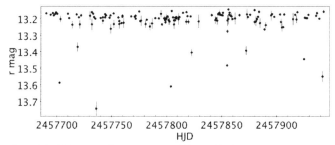

Figure 14. Light curve of GALEXASC J102502.22-294913.8.

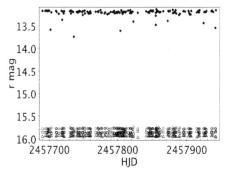

Figure 15. Light curve of GALEXASC J102502.22-294913.8 (solid circle) and its comparison stars (c1, open circle; c2, square; c3, triangle). Error bars are omitted.

Table 10. Cross-identifications of GALEXASC J102502.22-294913.8. Positional data were taken from NED and IRSA.

Identification	R. A. (2000) h m s	Dec. (2000) ° ′ ″
GALEXASC J102502.22-294913.8	10 25 02.22	–39 49 13.8
2MASS J10250224-3949140	10 25 02.241	–39 49 14.10
GALEXMSC J102502.26-394914.2	10 25 02.261	–39 49 14.23
USNO-B1.0 0501-0215578	10 25 02.24	–39 49 14.14

3.7. GALEXASC J102509.56-394736.7

This object was observed in the field of NGC 3244. It is classified as an infrared and ultraviolet source in NED. The cross-identifications of the object are listed in Table 11. Figure 16 is a light curve of the object containing data points from HJD 2457690 to 2457942. The brightness range over the observed period is ~0.5 magnitude. Figure 17 is a light curve of GALEXASC J102509.56-394736.7 and its comparison stars. Information about the comparison stars is listed in Table 8. The object shows a strong sign of variability in brightness in comparison to its comparison stars. The object exhibits an irregular pattern of variability in brightness, but as it is in the case of STSSL2 J041928.18-545750.7, further observations should be conducted of this target.

Table 11. Cross-identifications of GALEXASC J102509.56-394736.7. Positional data were taken from NED and IRSA.

Identification	R. A. (2000) h m s	Dec. (2000) ° ′ ″
GALEXASC J102509.56-394736.7	10 25 09.57	–39 47 36.7
2MASS J10250953-3947372	10 25 09.537	–39 49 47 37.22
GALEXMSC J102509.61-394737.2	10 25 09.	–39 49 47 37.3
USNO-B1.0 0502-0215192	10 25 09.61	–39 47 38.57

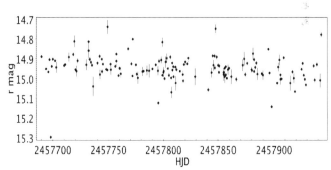

Figure 16. Light curve of GALEXASC J102509.56-394736.7.

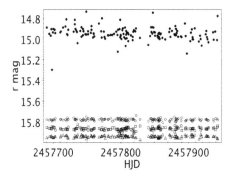

Figure 17. Light curve of GALEXASC J102509.56-394736.7 (solid circle) and its comparison stars (c1, open circle; c2, square; c3, triangle). Error bars are omitted.

3.8. SDSS J123801.86+115436.5

This star was observed in the direction of M58. It is also classified as an infrared source in NED. The cross-identifications of the object are listed in Table 12. Figure 18 is a light curve of SDSS J123801.86+115436.5 which was imaged from HJD 2457781 to 2457876. The brightness range over the observed period is ~0.15 magnitude. Figure 19 is a light curve of the object and its comparison stars. Table 13 lists information about the comparison stars. The brightness of the object increases around HJD 2457820. In comparison to its comparison stars, the variable star candidate shows a slightly stronger sign of variability in brightness. Because of the inconsistency and scarcity of this target's data, we were not able to conclude on the object's periodicity. Further observations should be conducted to clarify not only the periodicity of the object, but also its variability.

3.9. GALEXMSC J180742.51+174200.4

This object was observed in the field of NGC6555. It is classified as an ultraviolet source in IRSA. The cross-identifications of the object are listed in Table 14. Figure 20 is a light curve of GALEXMSC J180742.51+174200.4, which was imaged from HJD 2457864 to 2457960. IMSNG data of this field only covers a limited time span of 96 days. Figure 21 is a light curve of the variable star candidate and its comparison stars. Information about the comparison stars is listed in Table 15. The brightness range over the observed period is ~0.4 magnitude. The object shows strong signs of variability in brightness, but we were not able to identify any evidence of periodicity when running a period. As it is the case for SDSS J123801.86+115436.5, the object's variability behavior should not be concluded without further accumulation of observational data.

Table 12. Cross-identifications of SDSS J123801.86+115436.5. Positional data were taken from NED and IRSA.

Identification	R. A. (2000) h m s	Dec. (2000) ° ′ ″
SDSS J123801.86+115436.5	12 38 01.865	+11 54 36.54
2MASS J12380186+1154367	12 38 01.869	+11 54 26.72
SSTSL2 J123801.86+115436.8	12 38 01.867	+11 54 36.84
USNO-B1.0 1019-0238385	12 38 01.87	+11 54 36.59

Table 14. Cross-identifications of GALEXMSC J180742.51+174200.4. Positional were data taken from IRSA.

Identification	R. A. (2000) h m s	Dec. (2000) ° ′ ″
GALEXMSC J180742.51+174200.4	18 07 42.51	+17 42 00.5
2MASS 18074251+1742003	18 07 42.51	+17 42 00.37
USNO-B1.0 1077-0384323	18 07 42.53	+17 42 00.41

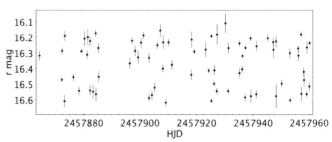

Figure 20. Light curve of GALEXMSC J180742.51+174200.4.

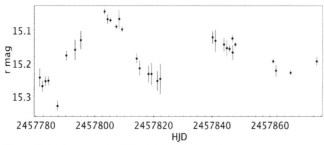

Figure 18. Light curve of SDSS J123801.86+115436.5.

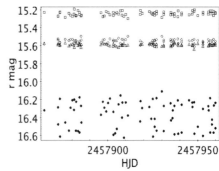

Figure 21. Light curve of GALEXMSC J180742.51+174200.4 (solid circle) and its comparison stars (c1, open circle; c2, square; c3, triangle). Error bars are omitted.

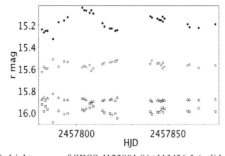

Figure 19. Light curve of SDSS J123801.86+115436.5 (solid circle) and its comparison stars (c1, open circle; c2, square; c3, triangle). Error bars are omitted.

Table 13. Comparison stars for SDSS J123801.86+115436.5. Data were taken from APASS.

Name[1]	Record No.[2]	R. A. (2000) h m s	Dec. (2000) ° ′ ″	r	r error
Comparison 1	27580451	12 37 29.9	+11 53 3.63	15.583	0.12
Comparison 2	27580450	12 38 14.97	+11 50 30.73	15.967	0.086
Comparison 3	27580456	12 37 48.52	+11 54 48.86	15.893	0.048

Notes: 1. in this paper; 2. in APASS.

Table 15. Comparison stars for variable star candidates in the field of NGC6555: GALEXMSC J180742.51+174200.4, GALEXASC J180759.71+173815.4, and 2MASS J18080948+1736557. Data were taken from APASS.

Name[1]	Record No.[2]	R. A. (2000) h m s	Dec. (2000) ° ′ ″	r	r error
Comparison 1	35377630	18 08 2.78	+17 41 37.74	15.511	0.02
Comparison 2	34985413	18 07 36.83	+17 35 32.33	15.275	0.048
Comparison 3	34985436	18 07 54.07	+17 37 12.55	15.609	0.163

Notes: 1. in this paper; 2. in APASS.

3.10. GALEXASC J180759.71+173815.4

This object was observed in the same field of GALEXMSC J180742.51+174200.4. It is classified as an infrared and ultraviolet source in NED. The cross-identifications of the object are listed in Table 16. Figure 22 is a light curve of GALEXASC J180759.71+173815.4 which was imaged from HJD 2457864 to 2457960. The brightness range over the observed period is ~0.3 magnitude. Figure 23 is a light curve of the variable star candidate and its comparison stars. Information about the comparison stars is listed in Table 15. Compared to its comparison stars, the GALEXASC J180759.71+173815.4 shows a stronger sign of variability in brightness. The object's variability in brightness appears to be irregular, and a period analysis failed to indicate any signs of periodicity. Additional observations would be beneficial to reveal more specific details about the object's pattern of variability.

Table 16. Cross-identifications of GALEXASC J180759.71+173815.4. Positional data were taken from IRSA.

Identification	R. A. (2000)	Dec. (2000)
	h m s	o ' "
GALEXASC J180759.71+173815.4	18 07 59.72	+17 38 15.4
2MASS 18075972+1738153	18 07 59.725	+17 38 15.39
USNO-B1.0 1076-0373592	18 07 59.71	+17 38 15.86

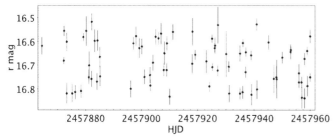

Figure 22. Light curve of GALEXASC J180759.71+173815.4.

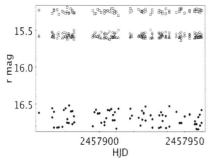

Figure 23. Light curve of GALEXASC J180759.71+173815.4 (solid circle) and its comparison stars (c1, open circle; c2, square; c3, triangle). Error bars are omitted.

3.11. 2MASS J18080948+1736557

This object was imaged in the field of NGC6555. The cross-identifications of the object are listed in Table 17. It is classified as an infrared and ultraviolet source in NED. Figure 24 is a light curve of 2MASS J18080948+1736557 which was imaged from HJD 2457864 to 2457960. Figure 25 is a light curve of the variable star candidate and its comparison stars. Information about the comparison stars is listed in Table 15. The brightness range over the observed period is ~0.4 magnitude. The light curve and period analysis of the object indicates that the object

is possibly an irregular variable, but further observations should be conducted as the data coverage of the target's variability behavior is limited.

Table 17. Cross-identifications of 2MASS J18080948+1736557. Positional data were taken from IRSA.

Identification	R. A. (2000)	Dec. (2000)
	h m s	o ' "
2MASS J18080948+1736557	18 08 09.49	+17 36 55.77
GALEXASC J18089.49+173656.2	18 08 09.49	+17 36 56.2
GALEXMSC J18089.48+173655.7	18 08 09.49	+17 36 55.8
USNO-B1.0 1076-0373717	18 08 09.49	+17 36 56.19

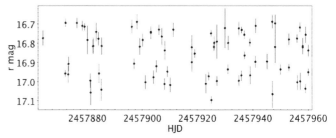

Figure 24. Light curve of 2MASS J18080948+1736557.

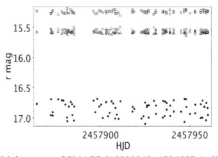

Figure 25. Light curve of 2MASS J18080948+1736557 (solid circle) and its comparison stars (c1, open circle; c2, square; c3, triangle). Error bars are omitted.

3.12. 2MASS J19422267-1019583

This object was imaged in the field of NGC 6814. It is classified as an infrared and ultraviolet source in NED. The cross-identifications of the object are listed in Table 18. Figure 26 is a light curve of 2MASS J19422267-1019583 which was imaged from HJD 2457872 to 2457960 for a total of 62 useful measurements. The brightness range over the observed period is ~0.4 magnitude. Figure 27 is a light curve of the variable star candidate and its comparison stars. Information about the comparison stars is listed in Table 19. The object shows a strong evidence of variability in brightness in comparison to its comparison stars. While the period analysis revealed no signs of periodicity, from a lack of data, we were not able to conclude whether the variable is irregular or periodic.

Table 18. Cross-identifications of 2MASS J19422267-1019583. Positional data were taken from NED and IRSA.

Identification	R. A. (2000)	Dec. (2000)
	h m s	o ' "
2MASS J19422267-1019583	19 42 22.673	–10 19 58.33
GALEXMSC J194222.66-101958.6	19 42 22.66	–10 19 58.6
GALEXASC J194222.70-101957.7	19 42 22.70	–10 19 57.8
USNO-B1.0 0796-0586608	19 42 22.70	–10 19 58.44

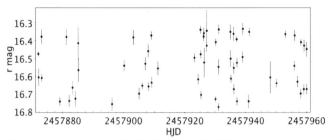

Figure 26. Light curve of 2MASS J19422267-1019583.

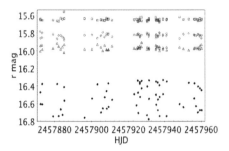

Figure 27. Light curve of 2MASS J19422267-1019583 (solid circle) and its comparison stars (c1, open circle; c2, square; c3, triangle). Error bars are omitted.

Table 19. Comparison stars for variable star candidates in the field of NGC6814: 2MASS J19422267-1019583, GALEXASC J194247.24-102215.3, and SSTSL2 J194231.40-102150.5. Data were taken from APASS.

Name[1]	Record No.[2]	R. A. (2000) h m s	Dec. (2000) ° ' "	r	r error
Comparison 1	30624988	19 42 43.66	−10 22 13.09	15.768	0.059
Comparison 2	30625086	19 42 35.97	−10 17 23.69	15.642	0.01
Comparison 3	30625364	19 42 28.53	−10 12 29.19	16.037	0.043

Notes: 1. in this paper; 2. in APASS.

3.13. GALEXASC J194247.24-102215.3

This object was imaged in the direction of NGC 6814. It is classified as an infrared and ultraviolet source in NED. The cross-identifications of the object are listed in Table 20. Figure 28 is a light curve of GALEXASC J194247.24-102215.3, which was imaged from HJD 2457872 to 2457960. Figure 29 is a light curve of the variable star candidate and its comparison stars. Information about the comparison stars is listed in Table 19. The brightness range over the observed period is ~0.3 magnitude. For the same reasons as the previous object, 2MASS J19422267-1019583, we were not able to conclude on the periodicity of GALEXASC J194247.24-102215.3.

Table 20. Cross-identifications of GALEXASC J194247.24-102215.3. Positional data were taken from NED and IRSA.

Identification	R. A. (2000) h m s	Dec. (2000) ° ' "
GALEXASC J194247.24-102215.3	19 42 47.24	−10 22 15.4
2MASS J19424724-1022151	19 42 47.250	−10 22 15.19
USNO-B1.0 0796-0586941	19 42 47.22	−10 22 15.54

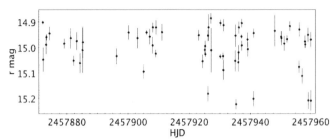

Figure 28. Light curve of GALEXASC J194247.24-102215.3. Error bars are omitted.

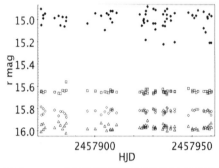

Figure 29. Light curve of GALEXASC J194247.24-102215.3 (solid circle) and its comparison stars (c1, open circle; c2, square; c3, triangle). Error bars are omitted.

3.14. SSTSL2 J194231.40-102150.5

This object is in the same field of view as the previous object. It is classified as an infrared source in NED. The cross-identifications of the object are listed in Table 21. Figure 30 is a light curve of SSTSL2 J194231.40-102150.5, which was imaged from HJD 2457872 to 2457960. The brightness range over the observed period is ~0.4 magnitude. Figure 31 is a light curve of the variable star candidate and its comparison stars. Information about the comparison stars is listed in Table 19. A period analysis revealed a possible period of 0.12497 days with no indication of aliases. The power spectrum is shown in Figures 32 and 33. Figure 34 is the phase plot of the object phased to a period of 0.12497 days.

Table 21. Cross-identifications of SSTSL2 J194231.40-102150.5. Positional data were taken from NED and IRSA.

Identification	R. A. (2000) h m s	Dec. (2000) ° ' "
SSTSL2 J194231.40-102150.5	19 42 31.409	−10 21 50.52
2MASS J19423139-1021503	19 42 31.394	−10 21 50.31
USNO-B1.0 0796-0586742	19 42 31.40	−10 21 50.49

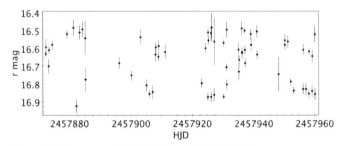

Figure 30. Light curve of SSTSL2 J194231.40-102150.5.

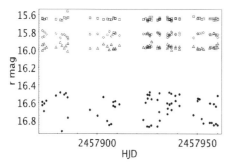

Figure 31. Light curve of SSTSL2 J194231.40-102150.5 (solid circle) and its comparison stars (c1, open circle; c2, square; c3, triangle). Error bars are omitted.

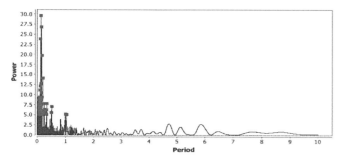

Figure 32. Power spectrum for SSTSL2 J194231.40-102150.5. Period is in days.

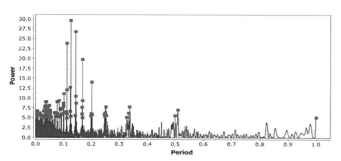

Figure 33. Power spectrum for SSTSL2 J194231.40-102150.5. Period is in days.

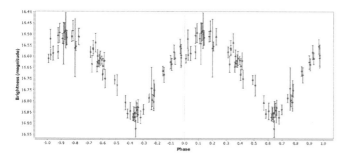

Figure 34. Phase plot of SSTSL2 J194231.40-102150.5 phased to a period of 0.12497 days. Brightness (vertical axis) is in r magnitude.

3.15. 2MASS J18180055-5443534

This object was imaged in the field of ESO182-G010. It is classified as an infrared source in NED. The cross-identifications of the object are listed in Table 22. Figure 35 is a light curve of 2MASS J18180055-5443534, which was imaged from HJD 2457864 to 2457960. Figure 36 is a light curve of the variable star candidate and its comparison stars. Information about the comparison stars is listed in Table 23. The variability of the object is evident; the brightness range over the observed period

Table 22. Cross-identifications of 2MASS J18180055-5443534. Positional data were taken from NED and IRSA.

Identification	R. A. (2000) h m s	Dec. (2000) ° ′ ″
2MASS J18180055-5443534	18 18 00.559	−54 43 53.20
USNO-B1.0 0352-0775609	18 18 00.54	−54 43 53.21

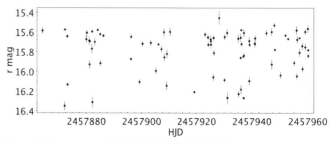

Figure 35. Light curve of 2MASS J18180055-5443534.

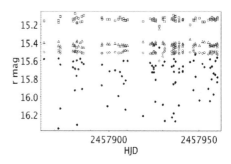

Figure 36. Light curve of 2MASS J18180055-5443534 (solid circle) and its comparison stars (c1, open circle; c2, square; c3, triangle). Error bars are omitted.

Table 23. Comparison stars for 2MASS J18180055-5443534. Data were taken from APASS.

Name[1]	Record No.[2]	R. A. (2000) h m s	Dec. (2000) ° ′ ″	r	r error
Comparison 1	58642670	18 18 19.02	−54 35 20.07	15.557	0.056
Comparison 2	58642652	18 17 53.64	−54 35 36.27	15.15	0.044
Comparison 3	58642371	18 18 21.25	−54 45 15.57	15.373	0.018

Notes: 1. in this paper; 2. in APASS.

is ~0.8 magnitude. Due to our sparse measurements, we were not able to reach a conclusion on the periodicity of the object. Further observations should be conducted on this target.

4. Conclusion

The results described here are preliminary; most variable sources discussed in this paper suffer from a lack of observational data, resulting in several open-ended conclusions on periodicity. Additional observations of these targets should be able to reveal their complete behavior of variability. As we have confirmed the objects' variabilities in brightness in this paper, the next step would be to uncover the cause of variability of these objects and categorize them into appropriate classes of variables. We would welcome the participation of AAVSO observers in this effort. We hope to make the photometric data and images of the survey available through a date archive in near future. In the meantime, interested researchers may contact the authors to obtain the data.

5. Acknowledgements

This work was supported by the National Research Foundation of Korea (NRF) grant, No. 2017R1A3A3001362. This research has made use of several resources including "Aladin sky atlas" and SIMBAD database operated at CDS, Strasbourg Observatory, France. We also used data from NASA/IPAC Extragalactic Database and NASA/IPAC Infrared Science Archive, which are operated by the Jet Propulsion Laboratory, California Institute of Technology, under contract with the National Aeronautics and Space Administration.

We thank the members of CEOU for their help and guidance dealing with IMSNG data. We are also grateful for the insightful guidance of Stella Kafka of American Association of Variable Star Observers, Cambridge, Massachusetts, and Caroline Odden of Phillips Academy, Andover, Massachusetts, throughout the research process.

References

Bertin, E., and Arnouts, S. 1996, *Astron. Astrophys., Suppl. Ser.*, **117**, 393.

Choi, C., and Im, M. 2017, *J. Korean Astron. Soc.*, **50**, 71.

Henden, A. A., Levine, S. E., Terrell, D., Smith, T. C., and Welch, D. 2012, *J. Amer. Assoc. Var. Star Obs.*, **40**, 430.

Im, M., Choi, C., and Kim, K. 2015b, *J. Korean Astron. Soc.*, **48**, 207.

Im, M., Choi, C., Yoon, S.-C., Kim, J.-W., Ehgamberdiev, S. A., Monard, L. A. G., and Sung, H.-I. 2015a, *Astrophys. J., Suppl. Ser.*, **221**, 22.

Jang, M., and Miller, H. R. 1997, *Astron. J.*, **114**, 565.

Kim, J., Karouzos, M., Im, M., Choi, C., Kim, D., Jun, H. D., Lee, J. H., Mezcua, M. 2018, *J. Korean Astron. Soc.*, **51**, 89.

Romero, G. E., Cellone, S. A., and Combi, J. A. 1999, *Astron. Astrophys., Suppl. Ser.*, **135**, 477.

New Observations, Period, and Classification of V552 Cassiopeiae

Emil J. L. Pellett

(Student at James Madison Memorial High School, Madison, Wisconsin), Yerkes Observatory, Williams Bay, WI 53191;
emil.pellett@live.com

Received October 9, 2018; revised November 1, 2018; accepted November 1, 2018

Abstract V552 Cas is a star yet to be systematically studied in the constellation Cassiopeia. The star was first indicated as a RR Lyr star by Götz and Wenzel (1956). The results presented in this paper convincingly demonstrate V552 Cas is a β Lyrae (EB) type eclipsing binary star with a period of 1.32808 days.

1. Introduction

Yerkes Observatory has a long history of astronomy education through outreach programs for both college and high school students. One of those programs, the McQuown Scholars program, allows high school students to conduct their own astronomical research through access to the Skynet system of robotic telescopes as well as the Stone Edge Observatory, with the support and supervision of experienced astronomers. The results presented in this paper were obtained as part of a McQuown Scholars project.

The variable star V552 Cas, located at R.A. $01^h 05^m 18.70^s$, Dec. +63° 21' 24.7" (2000), was discovered in 1956 as part of the Sonneberg Observatory surveys, but has not been systematically studied. Based on the original observations, V552 Cas was classified as a RR Lyrae (RRAB) type variable star in the *General Catalogue of Variable Stars* (GCVS; Samus *et al.* 2017). V552 Cas has been further studied by the WISE satellite and was found to be a RRAB variable star with a period of 0.64 ± 0.005 day (Gavrilchenko *et al.* 2014). The AAVSO International Database (AID) reports no observations of V552 Cas, and the AAVSO Photometric All Sky Survey (APASS; Henden *et al.* 2016) gives only two observations which show no variability.

V552 Cas was originally described in Götz and Wenzel (1956), who designated it as S 3873. Based on the observations taken from the few available plates at the time, the authors concluded that S 3873 was possibly a RR Lyrae type star, even though it had been first indicated as an Algol type star (see note in Götz and Wenzel 1956, p. 311). A more exact statement could not be made because insufficient data were available (Götz and Wenzel 1956). Based on this publication, the GCVS lists V552 Cas as a RRAB type variable star. A finder chart is provided by the Sonneberg Observatory (see http://www.4pisysteme.de/obs/pub/mvs/MVS_Volume_01.pdf page 291).

This report describes the first systematic study of V552 Cas. The presented results indicate V552 Cas should be classified as a β Lyrae (EB) type eclipsing binary star, not a RRAB type variable star.

2. Observations

Observations of V552 Cas were taken over 17 nights from August 2017 through February 2018 with a 20-inch (51-cm), f/8.1, Cassegrain telescope at Stone Edge Observatory in California (observatory code G52). The observations were made with a Finger Lakes Instrumentation PROLINE PL230 CCD camera. A binning of 2 × 2 was used which yielded a 26 by 26 arc-minute field of view and a 1.4 arc-second per pixel scale. The i, r, and g filters from the Sloan Digital Sky Survey's (SDSS) filter system were used for the observations.

A total of 2,576 images were obtained. Images with obvious defects (e.g., satellite trails through stars to be measured, light contamination from car headlights, variable cloud cover) were excluded from the final data set, leaving a total of 2,043 images that were analyzed. The images were processed using bias and flat frames taken on the same night and with dark frames taken by the observatory within the same week. The darks were scaled down from a 120-second exposure to the 70 seconds used for the exposures of V552 Cas.

Photometry of the processed images was performed with the SOURCE EXTRACTOR software (Bertin and Arnouts 1996) using a 5-pixel aperture radius. The default detection threshold of 1.5σ was sufficient to detect the required stars. The software calculated the sky background of the image. The full description of how SOURCE EXTRACTOR creates a background map can be found in section 7 of the user manual (see https://www.astromatic.net/pubsvn/software/sextractor/trunk/doc/sextractor.pdf). Differential magnitudes were calculated for V552 Cas and a check star against a comparison star of similar color to

Table 1. The position and color of the variable and comparison stars.

Star	R.A. (2000) h m s	Dec. (2000) ° ' "	Color (B–V)
V552 Cas	01 05 18.70	+63 21 24.700	0.760
2MASS 01050480+6320565 (check)	01 05 04.809	+63 20 56.576	0.939
2MASS 01053262+6322391 (comparison)	01 05 32.628	+63 22 39.148	0.792

V552 Cas. See Table 1 for details. The differential magnitudes were zero pointed using the magnitude of the comparison star as found in APASS (Henden *et al.* 2016). The CCD observations have been deposited in the AAVSO International Database under "V552 Cas."

3. Analysis

A preliminary look at the first night's light curve showed a time of nearly constant brightness followed by a deep drop. This suggested the star might be an eclipsing binary with a period of around one day rather than a RR Lyrae. Observations on further nights confirmed the eclipsing nature of the star. A period search on all the acquired data was done using the AAVSO VSTAR DC DFT task (Benn 2012), searching for periodicities in the range 0.2 to 2.0 days with a resolution of 0.00001 day. A clear periodicity of 0.66409 day was found, as shown in Figure 1. Taking into account the need for primary and secondary minima, which are of almost equal depth, the true period should be about twice the derived value.

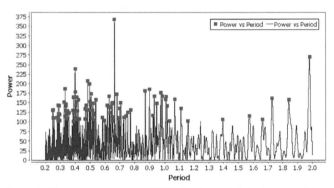

Figure 1. The power spectrum of V552 Cas for periods in the range 0.2 to 2.0 days.

A period of 1.32808 days was found to best fit all the data. Figure 2 shows the phased light curves plotted with this period for the three observed filters. The ephemeris used was:

Modified Julian Date (MJD) of Primary Eclipse in r-band = 57991.29571 + 1.32808 E.

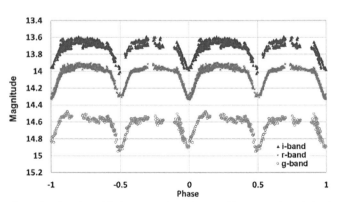

Figure 2. The phased light curve of V552 Cas with an Epoch (MJD) of 57991.29571.

Figure 3 shows the light curves for the check star, which are constant with a scatter of about 0.03 magnitude, which can be taken as the error of a single observation.

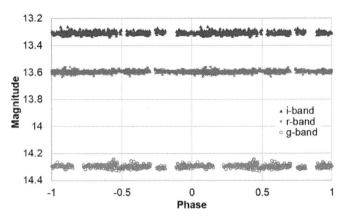

Figure 3. The phased light curve of the check star with the Epoch (MJD) of 57991.29571.

Examining the variable's light curves clearly shows that there are two minima of different depths and the more constant area is curved—not flat—showing V552 Cas is an EB type eclipsing binary star. Surprisingly, the deepest minima in the r data corresponds with the shallowest minima of the i and g data. No definitive explanation for this could be found. The eclipse depth of the deepest minima in the i filter is 0.37 magnitude, in the r filter 0.40, and in the g filter 0.36. It should also be noted that the out-of-eclipse light curves are not symmetrical, which is a strong sign of star spot activity, as is typical for RS CVn systems. This could be one possible explanation for the observed discrepancies in the eclipse depths in the SDSS i, r, and g filters. A detailed light curve analysis will be required to examine this hypothesis.

4. Summary and conclusions

Based on the presented data, V552 Cas has a variation of 0.45 magnitude in brightness and a period of 1.32808 days. V552 Cas should now be classified as an EB eclipsing binary star in accordance with the definitions presented by the GCVS.

5. Acknowledgements

A huge thank you is due to Dr. Wayne Osborn, the author's guide and mentor, without whom this work would have been impossible. A thank you is due to Yerkes Observatory and Kate Meredith for allowing use of its resources through the McQuown Scholars Program. The author also thanks Matt Nowinski, Amanda Pagul, Tyler Linder, and Marc Berthoud for their help and teaching.

References

Benn D. 2012, *J. Amer. Assoc. Var. Star Obs.*, **40**, 852.

Bertin, E., and Arnouts, S. 1996, *Astron. Astrophys., Suppl. Ser.*, **117**, 393.

Gavrilchenko, T., Klein, C. R., Bloom, J. S., and Richards, J. W. 2014, *Mon. Not. Roy. Astron. Soc.*, **441**, 715.

Götz, W., and Wenzel, W. 1956, *Veröff. Sternw. Sonneberg*, **2**, 284.

Henden, A. A., *et al.* 2016, AAVSO Photometric All-Sky Survey, data release 9 (http://www.aavso.org/apass).

Samus N. N., Kazarovets E. V., Durlevich O. V., Kireeva N. N., and Pastukhova E. N. 2017, *General Catalogue of Variable Stars*, version GCVS 5.1, in *Astro. Rep.*, **61**, 80 (http://www.sai.msu.su/gcvs/gcvs/index.htm).

Creating Music Based on Quantitative Data from Variable Stars

Cristian A. Droppelmann

Robarts Research Institute, Western University, 1151 Richmond Street, London, Ontario N6A 5B7, Canada; cdroppel@uwo.ca

Ronald E. Mennickent

Astronomy Department, Universidad de Concepción, Casilla 160-C, Concepción, Chile; rmennick@udec.cl

Received September 17, 2018; revised October 24, 2018; accepted November 21, 2018

Abstract   In this work we show a technique that allows for the musical interpretation of the brightness variations of stars. This method allows composers a lot of freedom to incorporate their own ideas into the score, based on the melodic line generated from the quantitative data obtained from the stars. There are a wide number of possible applications for this technique, including avant-garde music creation, teaching, and promotion of the association between music and science.

1. Introduction

Musical adaptation of the Universe can be understood as a scientific and artistic adventure. When we convert the changes of brightness of variable stars into music, two disciplines converge—astronomy allows us to detect, to record, and to interpret the changes in stars, and music allows us to represent those light fluctuations through art. This last step, entirely artistic, requires the creation of a method that allows us to analyze and interpret the energy flows received from stars using musical parameters, including pitch and rhythm. This activity requires intelligence and a good sense of aesthetics.

Few attempts of sound creation from stellar light curves have been made until now. Some examples of sonification—from the Kepler Special Mission—derived light curves from solar-type stars, red giants, cataclysmic variables, and eclipsing binary stars (NASA 2018). These sonifications have received the attention of the *New York Times* and were featured in an article entitled "Listening to the Stars" (Overbye 2011). Additionally, there are a few groups working in the sonification of diverse astronomical events, including gamma ray bursts (for a list of examples see Table 1). However, in the context of stellar light curves sonifications, they do not necessarily represent an artistic representation of the light curves, nor are they musically attractive. In this article, we propose a new method of converting star light curves into music, so that the resulting music would be more aesthetically pleasing to the human ear. This method is a new way of creating music and

has the potential to become a powerful tool for the development of avant-garde music, for musical teaching, for providing a listening experience to visually impaired people, based on astronomical data and phenomena, and for encouraging public interest in astronomy.

2. Composition

To create our first composition, we chose the star RV Tauri (R. A. (2000) 04^h 47^m 06.7^s, Dec. (2000) +26° 10' 45.6"), the prototype of the variable stars dubbed RV Tau stars, characterized by almost regular bright oscillations. These oscillations are characterized by alternating brightness minima that are modulated in irregular fashion. The AAVSO data available on-line (Kafka 2018) indicate that RV Tauri has a period of around 78 days and shows two maxima at V magnitude around 9.0, a minimum around magnitude 10.0, and another minimum about 0.5 magnitude fainter. This behavior cannot be fully explained, but is believed to be caused by chaotic stellar pulsations. Another reason we chose this star is because we had access to high-quality bright variation data that were collected over several years. The light curve was obtained from The *All Sky Automated Survey* catalog (Pojmański 1997). The curve consists of 180 photometric magnitudes in the V band, qualified as A-type (of the highest quality) obtained between Heliocentric Julian days 2452621.66922 and 2455162.70827, approximately seven years. The data are presented in two columns, Heliocentric Julian day versus

Table 1. Examples of astronomical data sonification.

Astronomical Data Sonification	Authors	Links
Exoplanets (discrete sounds and multiple instruments). Data sonification using Python MIDI-based code SONIFY.	Erin Braswell	https://osf.io/vgaxh http://astrosom.com/Aug2018.php
Gamma ray bursts (discrete sounds and multiple instruments).	Sylvia Zhu	https://blogs.nasa.gov/GLAST/2012/06/21/post_1340301006610/
Gamma ray bursts. Sonification based on ALMA astronomical spectra.	Tanmoy Laskar	https://public.nrao.edu/the-sound-of-one-star-crashing-haunting-melody-from-the-death-of-a-star/
Flaring Blazar (Several techniques).	Matt Russo Andrew Santaguida	http://www.system-sounds.com/the-creators/ https://svs.gsfc.nasa.gov/12994

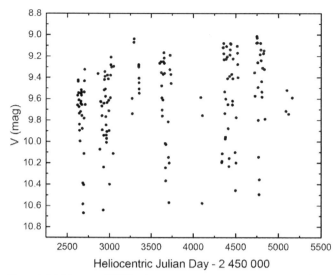

Figure 1. Light curve of RV Tauri.

V magnitude. The curve, presented in Figure 1, shows the light change of the star versus time.

To create the musical composition, the magnitudes of the star were converted into musical notes using the following equation:

$$M_N = \frac{(M - M_{min})}{(M_{max} - M_{min})} \times S \qquad (1)$$

Where: M_N = Normalized magnitude; M_{min} = minimal magnitude; M_{max} = maximal magnitude; M = magnitude in the chosen Heliocentric Julian day; S = Number of semitones—in this case 24 semitones (2 octaves).

As a consequence, one unit of normalized magnitude corresponds to one semitone. We assigned only 24 semitones to avoid excessive chromaticism; however, this parameter could be changed by a composer to include more chromaticism or include microtonality, if desired. Also, the composer could choose between using high magnitude values to higher tones or

vice versa only by assigning the higher normalized value to the higher note of the 24 semitones or the lower normalized value to the higher tone (see Table 2). For artistic reasons in this work we used higher normalized values to higher tones (Option 1 in Table 2).

A preliminary tone assignment for the RV Tauri data was created following the information of Table 1. Then, after the definition of the rhythm for the music, it was assigned a key with 3 sharp alterations (A major) that fitted the chromaticism of the melodic line generated by the stellar information, while keeping the pitches of the notes. Then, considering artistic reasons (to suit the capabilities and range of a string orchestra) the melodic line was transposed to E major (4 sharps) and finally, accidentals were eliminated enharmonically to simplify the interpretation by musicians. Consequently, the range of the stellar melodic line in the final score goes from G4 to G6.

The rhythm was assigned based on the time interval between magnitude measurements in Heliocentric Julian days using the following equation:

$$t_N = t_n - t_{n-1} \qquad (2)$$

Where: t_N = Normalized time interval; t_n = time of selected magnitude measurement in Heliocentric Julian days; t_{n-1} = time of previous magnitude measurement in Heliocentric Julian days. Note: The first Heliocentric Julian day has assigned a normalized value of 1.

Once we obtained the normalized time interval, we assigned the rhythm based on a range that goes from eighth notes to whole notes (see Table 3). This method allows for the creation of a rich and diverse rhythm pattern for the score, including the incorporation of intermediate-length musical notes (dotted notes) under the composer-determined criteria when normalized time interval values are close to the next assignation interval. Any gap in the brightness measurement created longer notes. However, the creation of rests is under control of the composer. In the music conceived for this paper no rests were used since our goal was to produce a feeling of continuity in the music. This criterion was used because a variable star only has changes in its brightness and not interruptions.

Table 2. Assignment of tones after normalization of magnitude values using a 24 semitones chromatic scale.

Tone (chromatic scale)	M_N Option 1 (higher note = higher mag. value)		M_N Option 2 (higher note = lower mag. value)	
C	0 to 0.99	(octave higher)	24	(octave higher)
C# / Db	1 to 1.99	13 to 13.99	23 to 23.99	11 to 11.99
D	2 to 2.99	14 to 14.99	22 to 22.99	10 to 10.99
D# / Eb	3 to 3.99	15 to 15.99	21 to 21.99	9 to 9.99
E	4 to 4.99	16 to 16.99	20 to 20.99	8 to 8.99
F	5 to 5.99	17 to 17.99	19 to 19.99	7 to 7.99
F# / Gb	6 to 6.99	18 to 18.99	18 to 18.99	6 to 6.99
G	7 to 7.99	19 to 19.99	17 to 17.99	5 to 5.99
G# / Ab	8 to 8.99	20 to 20.99	16 to 16.99	4 to 4.99
A	9 to 9.99	21 to 21.99	15 to 15.99	3 to 3.99
A# / Bb	10 to 10.99	22 to 22.99	14 to 14.99	2 to 2.99
B	11 to 11.99	23 to 23.99	13 to 13.99	1 to 1.99
C	12 to 12.99	24	12 to 12.99	0 to 0.99

Table 3. Assignment of rhythm after normalization of time intervals between Heliocentric Julian days for RV Tauri data.

Normalized Time Interval	Note Assigned
0 to 0.99	Eighth note
1 to 3.99	Quarter note
4 to 15.99	Half note
16 to 63.99	Whole note
Over 64	Ligated whole notes

Note: If more rhythmic richness is desired, the assignment can be initiated from thirty-second or sixteenth note. Also, a specific interval for dotted notes could be assigned.

Once the melodic line from the stellar data was created (Figure 2), the orchestral arrangement started with a pedal note based on the tonic note of the key used for the music (E major), and was structured as a four-voice canon with a major sixth interval. It should be noted that at this stage, the composer has complete freedom to arrange the composition based on the stellar melodic line. The score present here is only one of infinite possibilities and denotes the richness and potential of this method for the creation of new music, where even one star allows for the creation of an immense range of new music.

The score was made for a string orchestra, including first and second violins, violas, cellos, and double basses. The tempo was set at 50 bpm, and used simple time signature (4/4), equal temperament, and baroque pitch (A = 415 Hz). Finale 25 software and the Garritan library were used to create the score and the audio file. The music starts with the pedal note being played by double basses. Then, the melodic line (the music derived from the star light curve) is introduced by the first violins in the third measure. In the fourth measure, the second violins begin the canon as the second voice. Then, the violas begin in the fifth measure and finally, the cellos in the sixth measure. From there, the musical body is developed by following the melodic line derived from the star, until the last musical note corresponds to the last magnitude measured. At this point, the canon is ended to give a sense of completion, as is necessary for a musical score (Figure 3). The full score and the audio file can be downloaded as supplementary files.

3. Conclusions

Here, we have developed a technique that allows for the musical interpretation of the brightness variations of stars. The advantage of this method is that it allows composers a lot of freedom to incorporate their own ideas into a score, based on the melodic line obtained from stars. There are a wide number of possible applications for this technique, including avant-garde music creation, teaching, and promotion of the association between music and science. It could even be attractive

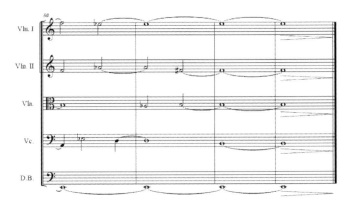

Figure 2. First 28 measures of the melodic line generated for RV Tauri.

Figure 3. Last 4 measures of the orchestral score based in the melodic line generated for RV Tauri.

for persons with visual impairments, who are interested in astronomy, to "hear" the universe rather than "see" it.

References

Kafka, S. 2018, variable star observations from the AAVSO International Database (https://www.aavso.org/aavso-international-database).

NASA. 2018, Kepler soundcloud (https://soundcloud.com/nasa/kepler-star-kic7671081b-light-curve-waves-to-sound)

Overbye, D. 2011, *New York Times* (January 30; http://www.nytimes.com/2011/01/31/science/space/31star.html?_r=2).

Pojmański, G. 1997, *Acta Astron.*, **47**, 467.

Supplementary information

Link to music (for listening or download):

https://my.pcloud.com/publink/show?code=XZDua17ZafxBcBqhi2HOQgi0kVlR0uh6TPky

Singular Spectrum Analysis: Illustrated by Application to S Persei and RZ Cassiopeiae

Geoff B. Chaplin

Hokkaido, Kamikawa-gun, Biei-cho, Aza-omura Okubo-kyosei, 071-0216, Japan;geoff@geoffgallery.net

Received June 27, 2018; revised September 24, October 10, 2018; accepted October 11, 2018

Abstract   We describe two methods of singular spectrum analysis, a data driven technique, providing and using code to analyze example data series, and introducing the public domain R package "Rssa." The analysis provides potential information about the underlying behavior of the series, stripping out noise, and is a pre-requisite for some further work such as non-linear time series analysis. Examples are taken from a long time series of S Per magnitude observations, and secular period changes in, and high frequency magnitude variations of, RZ Cas.

1. Introduction

Singular Spectrum Analysis ("SSA") has gained popularity since the mid-1980s as a data driven rather than model driven method for the analysis of time series in a wide range of disciplines, from meteorology, to medical sciences, engineering, finance, and physics. Papers on SSA commenced with Broomhead and King (1986a, 1986b), although some ideas can be traced back before this. Other influential early papers are Fraedrich (1986), Vautard and Ghil (1989), Vautard *et al.* (1992), and Allen and Smith (1996). Development of SSA was paralleled by Danilov and Zhigljavsky (1997) in the former USSR. References to recent publications in a wide range of different areas can be found in Zhigljavsky (2010) and by various authors in two issues of *Statistics and its Interface* (2010, 2017). The review paper by Ghil *et al.* (2002) gives an extensive list of references to earlier work.

The descriptions given here follow Golyandina *et al.* (2001), Golyandina and Zhigljavsky (2013), and Golyandina *et al.* (2018) which additionally provide many further examples from a variety of disciplines; the last additionally gives an extensive list of related research articles. SSA provides inter alia a means of identifying three major components of the time series to be analyzed. Long-term variations ("trends") form a component which typically is difficult to predict without some form of model or understanding of the underlying process. Fourier analysis usually reflects this long-term behavior as a rise in amplitude as frequency decreases. Noise elements are those with no particular pattern, generally weak, and often related to the observation process. Most importantly we wish to extract the "signal," manifested a some form of periodic variation, as both an end in itself and a pre-requisite for further analysis, for example non-linear time series analysis (NLTSA) or where further analysis requires a stationary data series. NLTSA has been used, albeit based on different techniques, in astronomical literature by Kollath (1990) and Buchler *et al.* (1996), and others in the context of giant variable stars. Methodology for NLTSA based on SSA analysis, and code, is provided by Huffaker *et al.* (2017) and code therein forms the basis of the code in this paper.

We describe in detail two methods of SSA along with code implementing these approaches. The code in Appendix A.1 and A.2 closely follows the mathematical methods, while the code in Appendix A.3 calls more efficient (but black box) library code. We illustrate the methods by application to two long-term astronomical time series—visual observations of the magnitude of the semi-regular variable S Per, and observed minus calculated times of minimum for the eclipsing binary RZ Cas, together with an analysis of high-frequency CCD/DSLR magnitude observations stripping out a signal which is far weaker than the noise in the data and revealing δ Scuti-type variations.

In this paper we use the R (R Foundation 2018a) statistical programming language and CRAN (R Foundation 2018b) libraries and in particular the function "ssa" in the R library "Rssa" (R Foundation 2018c). In the Appendix we provide code adapted from Huffaker *et al.* (2017) to perform the analysis. RStudio (2018) provides a convenient user interface to the R code and many of the charts below are taken directly from the RStudio platform.

2. Methodology

We use two different approaches ("1d-ssa" and "Toeplitz-ssa") to the construction of the "trajectory matrix" and the "lagged correlation matrix" after which reconstruction of the series follows the same process.

2.1. Decomposition—"1d-ssa"

A data series taken at equal time points is first adjusted by removing the average value and may then be represented by a set of numbers $(O_1, O_2, \dots, O_n)$ where n is 900 for example. A first column $(O_1, O_2, \dots, O_k)$, a second column $(O_2, O_3, \dots, O_k, O_{k+1})$ and a third column $(O_3, O_4, \dots, O_k, O_{k+1}, O_{k+2})$ (and so on) can rather trivially be produced starting one observation later (a "delay" on one) for some k < n. Each of these rows is called a "lagged" or "Takens" vector and stacking *m* (for example, 400) such columns next to each other produces what is termed the "trajectory matrix," X, where in our example X has 400 columns and $k = 501$ rows ($k = n - m + 1$ so that all the data are used). In this paper we take m to be a little under half the length of the time series; general advice is that m should be sufficiently large that we capture the main features of the data but less than half the length of the time series. In addition if a strong periodic signal is present m should be a multiple of the period. Golyandina and Zhigljavsky (2013) gives many examples where some choices of the embedding dimension are however very different from half the length of the series. Step 1 in the "SVD code" in the Appendix creates the trajectory matrix after reading in the data. (The data file should be formatted in column(s) as a csv file with the first row naming the column(s).)

The transpose of the trajectory matrix multiplied by the matrix gives an $m \times m$ matrix, $S = X^TX$, whose terms are covariances of the observations and is called the "lagged correlation matrix" and where m is called the "embedding dimension." Step 2 creates the lagged correlation matrix, and code is given in Appendix A.1.

2.2. Decomposition—"Tocplitz-ssa"

The series must be approximately stationary for Toeplitz decomposition (Golyandina and Zhigljavsky (2013) section 2.5.3). The first column of the trajectory matrix is the entire series, the second column as above starts at O_2 but pads the end with zero, the third column starts at O_3 and has two zeros at the end and so on. The lagged correlation matrix is calculated not as above but from the formula

$$S_{ij} = \sum_{t=1}^{t=n-|i-j|} O_t \, O_{t-|i-j|} / (n-|i-j|) \qquad (1)$$

The lagged correlation matrix again has m eigenvalues and eigenvectors. The alternative code is given in Appendix A.2.

2.3. Singular value decomposition—"SVD"

Any matrix of the form of S has m eigenvectors (see, for example, Lang 2013), $EV1 <= i <= m$ such that EVi multiplied by S simply stretches the EVi by a factor Li (the "eigenvalue") but doesn't change its direction. These eigenvectors also have the property that they are perpendicular to each other so define axes in m-dimensional space. We sort the eigenvectors in order from strongest eigenvalue to the weakest. The vectors

$$V_i = XE_i / \sqrt{L_i} \qquad (2)$$

(the eigenvalue term being introduced merely for normalization) are a projection of the time series of observations onto that eigenvector axis. The relative strength associated with each component is L_i / L where L is the sum of the eigenvalues. Step 3 performs these calculations and in our example under 1d-ssa V is a 501-length vector, whereas under Toeplitz it has length 900. Step 3 also writes the eigenvalues to a file and plots the relative magnitudes on a log scale.

The trajectory matrix decomposes into $X = X_1 + ... X_m$ where

$$X_i = V_i E_i^T \sqrt{L_i} \qquad (3)$$

and (each X_i has rank 1 and), $[\sqrt{L_i}, E_i, V_i]$ is referred to as the ith eigentriple of the SVD of X. Step 4 calculates the decomposition.

2.4. Series reconstruction

The values in each X_i are then averaged across "anti-diagonals" (row + column = constant) to give a time series component $\{x_i\}$ of the signal where in both decomposition cases the component has length n (900 in our example). Note that in the case of 1d-ssa decomposition the averaging is over 400 values for $400 <= i <= 500$, whereas under Toeplitz

the averaging stops at row 900 of each X. Step 5 of the "SVD code" calculates the individual averaged series, produces a graphic of the correlations between the m different time series, produces a graphic of a portion of the time series, and writes the series to a data file. The user selects how many series to plot in the user input section—typically starting with 40 or so then refining to 10 or 20.

The "reconstructed signal(s)" we choose for further analysis is a sum of a subset of the component signals where the signals meet certain requirements—not being part of the noise, having similar periodicity (trend or high frequency variation) and being sufficiently independent from other signals, as illustrated below. The graphical results help in this decision making process: time series which are highly correlated should be grouped together, time series which have no correlation with other signals can be treated as a separate signal; time series with very different periodicities / patterns may be better treated separately.

The more efficient Rssa package can be used instead of the above code to perform the same calculations and graphical analysis, and is also illustrated in Appendix A.3.

3. S Per magnitude variability

S Per (GSC 03698-03073) is an M4.5-7Iae C spectral type (Wenger *et al.* 2000) Src (Kiss *et al.* 2006) variable star with period(s) variously identified as 813 ± 60 (Kiss *et al.* 2006); 822 (Samus *et al.* 2017); 745, 797, 952, 2857 (Chipps *et al.* 2004). The strong color causes significant differences in the estimation of magnitude by visual observers arising from observer dependent color response, the Purkinje effect (Purkinje 1825), local atmospheric conditions, altitude of the star at the time of observation and other factors, giving rise to a significant level of noise related to the observation process—"extrinsic" noise. In addition there may be "intrinsic" random variability caused by the star and its environment—for example, matter thrown off by the star may form a non-uniform cloud causing variation over time in scattering of the light away from the observer. Data are taken from the BAA (2018) and the AAVSO (Kafka 2018) databases, and from the VSOLJ (2018) database prior to 2000. We restrict our attention to observations made by experienced observers (defined as those reporting over 100 observations of the star).

Figure 1a shows visually estimated magnitudes from experienced observers starting from JD 2423000 grouped into 878 40-day buckets. The buckets contained two empty buckets and values were estimated by linear interpolation from neighboring observations. Had the number of missing points been substantial, then a more sophisticated interpolation method, such as in Kondrashev and Ghil (2006), would be appropriate.

Applying 1d-ssa analysis Figure 1b shows a sharp drop in strength after the fourth eigenvector and another after the twelfth EV. A "scree test" (Cattell 1965a, 1965b) is often used to decide where signal ends and noise starts but in the case of very noisy data (for example visual magnitude observations of narrow range red variables) there may nevertheless be an uncorrelated but weak signal present after the strong presence of the noise begins and such a signal should not be ignored. Figure

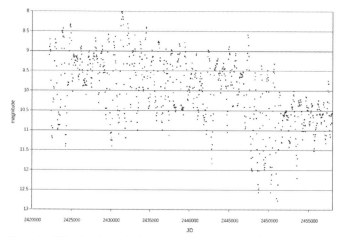

Figure 1a. S Per visual magnitude estimates by experienced observers averaged in 40-day buckets.

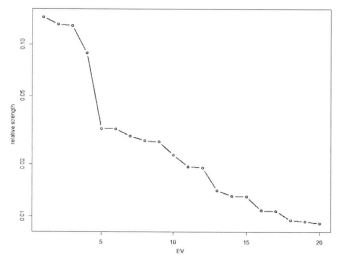

Figure 1b. S Per EV norms using 1d-ssa.

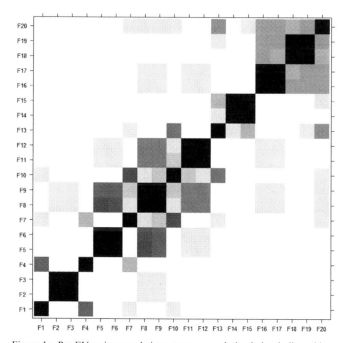

Figure 1c. Per EV series correlations, strong correlation being indicated by a more solid shade.

1c shows EVs 1 and 4 strongly correlated with each other but not the rest; EVs 2 and 3 strongly correlated but again not with the rest; EVs 7, 10, and 13 are largely detached from the rest while remaining EVs 5 to 12 form a block. Figure 1d shows similar behavior in EVs 1 and 4, and similar behavior in EVs 2 and 3, and EVs 5, 6, 8, 9, 11, 12.

Figure 1e shows the data together with the reconstructed signals from EVs 1 and 4, and from EVs 2 and 3 (the mean magnitude has been added back to these series). It is clear that EVs 1 and 4 (dashed line) reflect long-term trends (in particular coping with a shift in magnitude described below) while EVs 2 and 3 represent a 799-day oscillation (calculated separately). The figure shows (solid line) the reconstructed signal from EVs 2,3,5,6,8,9,11,12 which shows virtually the same periodicity as EVs 2 and 3 alone.

It is manifestly clear that the time series is non-stationary, there being a marked fall in brightness starting around 2447000 and being maintained. The relatively abrupt change in magnitude is discussed in Chipps *et al.* (2004), and Sabin and Zijlstra (2006) identify similar abrupt changes in other long-period variable stars. We make the following adjustment in order to produce a time-series which is closer to a stationary one. The adjusted magnitude at time t, m_t, is given by

$$m_t = raw_t - (t - T_t) \times H_t - K_t \qquad (4)$$

where *raw* is the observed magnitude and the parameters H and

Table 1. H and K parameters.

Time Period (JD)	H	K
< 2447000	0	0
2447000 to 2448750	4.87–04	0
2448750 to 2458093 (end)	0	0.853

K are given in Table 1.

Figure 1f shows the same data series as Figure 1a after adjustment described above.

Applying Toeplitz decomposition Figure 1g shows a clearer distinction between different eigenvectors, and following similar logic to above we group EVs 1–4, and 5–7 and the reconstructed series are shown in Figure 1h. EV 1–4 has a strong period at 815 days.

4. RZ Cas period variability and δ Scuti variation

4.1. Period variability

RZ Cas (GSC 04317-01793) is a semi-detached Algol-type binary comprising a primary A3V star (Duerbeck and Hänel 1979) and a carbon (Abt and Morrell 1995) or K01V (Maxted *et al.* 1994; Rodriguez *et al.* 2004) star which fills its Roche lobe. Times of minimum (tmin) were taken from the Lichtenknecker database (Frank and Lichtenknecker 1987) and compared with expected times using a linear ephemeris and a period of 1.19525031 days chosen to minimize the variance of the differences of observed tmin minus calculated tmin (O–C). Data was bucketed into 100-day blocks with a small number of missing values linearly interpolated between neighboring values

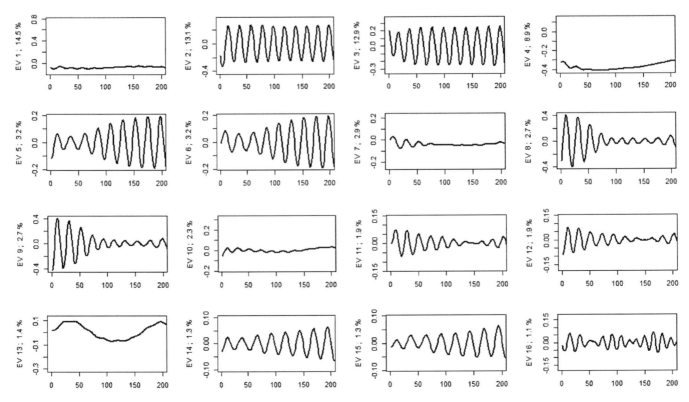

Figure 1d. S Per 1d-ssa first 16 individual EV time series, initial 200 data points – to identify the broad type of pattern.

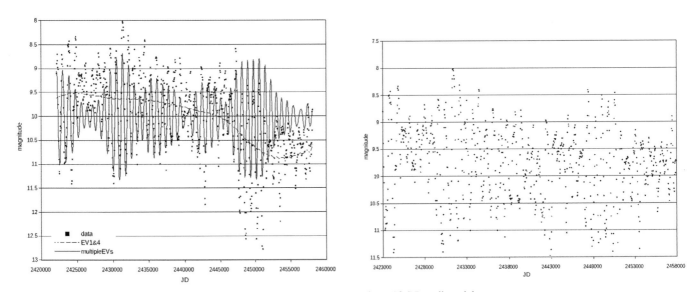

Figure 1e. S Per unadjusted data and reconstructed signals from EV groups using 1d-ssa.

Figure 1f. S Per adjusted data.

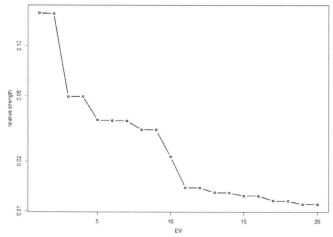

Figure 1g. S Per EV norms using Toeplitz decomposition.

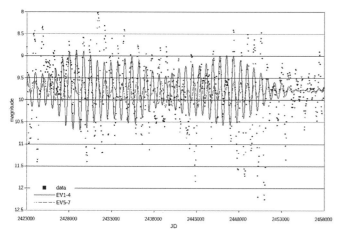

Figure 1h. S Per adjusted data and reconstructed signals from EV groups.

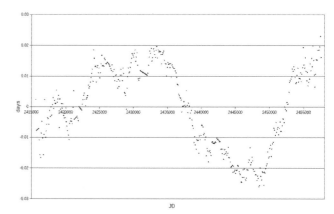

Figure 2a. RZ Cas O-C bucketed into 100-day intervals.

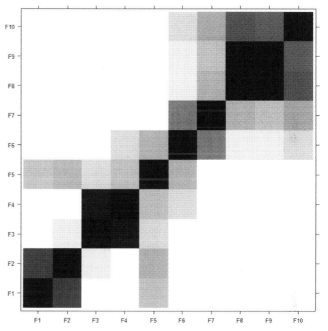

Figure 2b. RZ Cas O-C first 10 EV correlations.

and is shown in Figure 2a.

1d-ssa decomposition was applied, and inspection showed a typical noise pattern after EV6. The correlation matrix and time series charts (and eigenvalue magnitudes) and are shown in Figures 2b (first 10 EVs) and 2c (first 9 EVs).

From these two charts we see EVs 1 and 2 have a small correlation with EV 5 and similar periodicity (and comprise over 90% of the data variation), EVs 3 and 4 are largely separate (comprising 5%) and a similar period but shorter than EVs 1 and 2, EVs 6 and 7 (and several others also with significant correlation) seem to be correcting for abnormalities at the start of the period, and EV 6 and beyond may be regarded as the noise. Figure 2d shows the data and reconstructed signals.

The signal EV3–4 is intriguing: it should be borne in mind that the reconstructed signal is merely an average of the original data series (albeit a very complicated one)—at no point are harmonics used in the calculation, yet this signal is at first glance similar to a sine wave with period just of approximately 23 years. A closer look shows the amplitude of the signal is decreasing and the wavelength is not constant, although this could be a corruption caused by the original noisy data. Furthermore it is known (for example Allen and Smith 1996; Greco *et al.* 2015) that noise other than white noise—in particular noise related to an autoregressive process—can generate spurious periodicities. Further testing, which is beyond the scope of the current paper,

is required to determine whether the observed signal has arisen by chance or from a more complicated underlying non-linear process. If this was indeed harmonic and caused by an orbiting third body then the semi-amplitude of the signal implies an orbit for the main components about a center of mass of the system, and the joint masses of the eclipsing stars would imply a mass of under 0.2 solar mass for an orbiting body.

4.2. δ Scuti-type variation

High frequency CCD observations by G. Samolyk from the AAVSO database were analyzed as follows. Differences from model magnitudes (using a Wilson-Devinney eclipse model (see, for example, Kallrath and Milone 2009)) were calculated and analyzed using 1d-ssa decomposition (Figures 3a and 3b). EVs1–4 and 7, 8 represent the slow deviations, and a relatively strong EV5–6 is independent of other signals apart from the weak 14 and 15, and shows a clear periodicity of 22.4 minutes—in good agreement with Ohshima *et al.* (2001) and Rodriguez *et al.* (2004). The very high frequency variations in EVs 9–12 may be instrumentation related. Figure 3c shows the

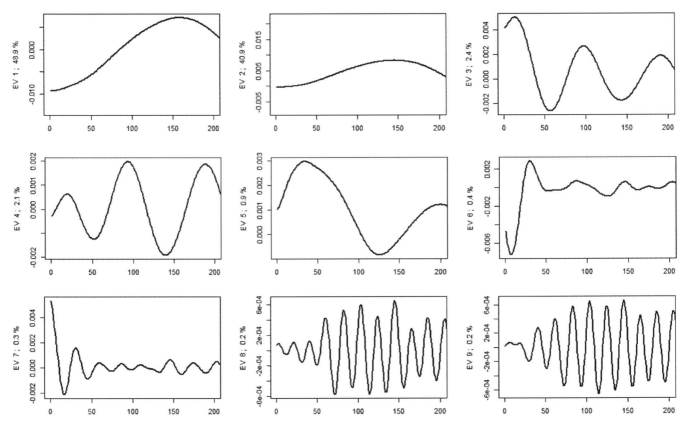

Figure 2c. RZ Cas first 9 individual EV time series, initial 200 data points—to identify the broad type of pattern.

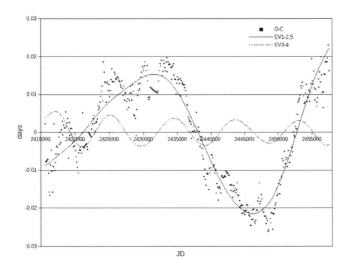

Figure 2d. RZ Cas O–C and reconstructed signals from EV groups.

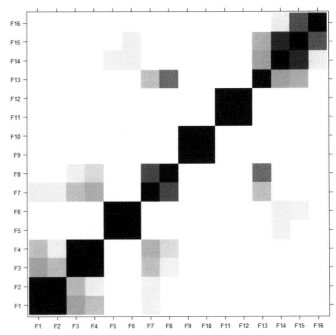

Figure 3a. RZ Cas high-frequency CCD data, EV correlation matrix, first 16 EVs.

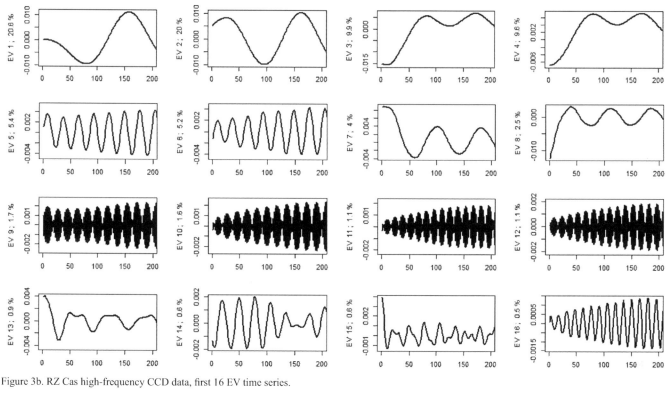

Figure 3b. RZ Cas high-frequency CCD data, first 16 EV time series.

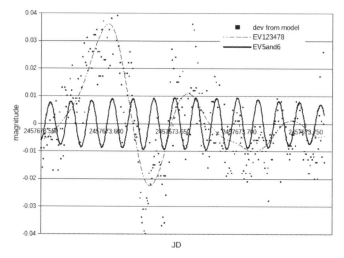

Figure 3c. RZ Cas high-frequency CCD data and reconstructed signal from EV groups.

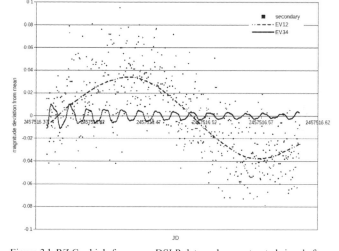

Figure 3d. RZ Cas high-frequency DSLR data and reconstructed signals from EV groups.

data and reconstructed signal.

A second series of relatively noisy DSLR data shows virtually the same periodicity—EV34 shows periodicity of 21.1 minutes. DSLR data by Screech from the BAA database taken through a secondary minimum have been analyzed simply by removing the mean then applying 1d-ssa decomposition and analysis, the result being shown in Figure 3d.

5. Conclusions

In this paper we have given code corresponding closely to the formulae behind singular spectrum analysis as well as code to call the corresponding more efficient "black box"

functionality in the Rssa R code package. We have shown how to use the key intermediate output—eigenseries correlation plots and eigenseries time series plots together with their relative strengths—to reconstruct meaningful time series components—trend, periodic and residual noise—of the original time series. As an interactive data driven method this is more revealing, and capable of extracting more information, that typical model driven methods. The S Per data example is one of simple periodicity discovery and is included to illustrate SSA and its application. The RZ Cas O–C series discovers a periodic signal in the times of minimum and hints at a possible third body, while the high frequency data illustrate how the δ Scuti variability can be extracted from relatively noisy data exhibiting strong long-term variation throughout the data sample.

6. Acknowledgements

The author is grateful to the AAVSO, the BAA, and the VSOLJ for providing the observational data used in this analysis. The author would also like to thank Prof. Ray Huffaker for helpful suggestions and encouragement during the production of this paper, and an anonymous referee whose suggestions significantly improved the paper.

References

Abt, H. A., and Morrell, N. 1995, *Astrophys. J., Suppl. Ser.*, **99**, 135.

Allen, M. R., and Smith, L. A. 1996, J. *Climate*, **9**, 3373.

British Astronomical Association Variable Star Section. 2018, BAAVSS online database (http://www.britastro.org/vssdb/).

Broomhead, D. S., and King, G. P. 1986a, *Phys. D: Nonlinear Phenomena.* **20**, 217.

Broomhead, D. S., and King, G. P. 1986b, in *Nonlinear Phenomena and Chaos*, ed. S. Sakar, Adam Hilger, Bristol, 113.

Buchler, J. R., Kollath, Z., Serre, T., and Mattei, J. 1996, *Astrophys. J.*, **462**, 489.

Cattell, R. B. 1965a, *Biometrics*, **21**, 190,

Cattell, R. B. 1965b, *Biometrics*, **21**, 405.

Chipps, K. A., Stencel, R. E., and Mattei, J. A. 2004, *J. Amer. Assoc. Var. Star Obs.*, **32**, 1.

Danilov, D. and Zhigljavsky, A., eds. 1997, *Principal Components of Time Series: the "Caterpillar" Method*, St. Petersburg Univ. Press, Saint Petersburg (in Russian).

Duerbeck, H. W., and Hänel, A. 1979, *Astron. Astrophys. Suppl. Ser.*, **38**, 155.

Fraedrich, K. 1986, *J. Atmos. Sci.*, **4**, 419.

Frank, P. and Lichtenknecer, D. 1987, *BAV Mitt.*, No. 47, 1 (Lichtenknecker database http://www.bav-astro.eu/index. php/veroeffentlichungen/service-for-scientists/lkdb-engl).

Ghil, M., *et al.* 2002, *Rev. Geophys.*, **40**, 1.

Golyandina, N., Korobeynikov, A., and Zhigljavsky, A. 2018, *Singular Spectrum Analysis with R*, Springer-Verlag, Berlin, Heidelberg.

Golyandina, N., Nekrutkin, V., and Zhigljavsky, A. 2001, *Analysis of Time Series Structure*, CRC Press, Boca Raton, FL.

Golyandina, N., and Zhigljavsky, A. 2013, *Singular Spectrum Analysis for Time Series*, Springer-Verlag, Berlin, Heidelberg.

Greco, G., *et al.* 2015, *Astrophys. Space Sci. Proc.*, **42**, 105.

Huffaker, R., Bittelli, M., and Rosa, R. 2017, *Non-linear Time Series Analysis with R*, Oxford Univ. Press, Oxford.

Kafka, S. 2018, variable star observations from the AAVSO International Database (https://www.aavso.org/aavso-international-database).

Kallrath, J., and Milone, E. F. 2009, *Eclipsing Binary Stars: Modeling and Analysis*, Springer-Verlag, Berlin, Heidelberg.

Kiss, L. L., Szabo, Gy. M., and Bedding, T. R. 2006, *Mon. Not. Roy. Astron. Soc.*, **372**, 1721.

Kollath, Z. 1990, *Mon. Not. Roy. Astron. Soc.*, **247**, 377.

Kondrashov, D., and Ghil, M. 2006, *Geophys.*, **13**, 151.

Lang, S. 2013, *Linear Algebra*, 3rd ed., Springer-Verlag, Berlin, Heidelberg.

Maxted, P. F. L., Hill, G., and Hilditch, R. W. 1994, *Astron. Astrophys.*, **282**, 821.

Ohshima, O., *et al.* 2001, *Astron. J.*, **122**, 418.

Purkinje, J. E. 1825, *Neue Beiträge zur Kenntniss des Sehens in Subjectiver Hinsicht*, Reimer, Berlin, 109.

The R Foundation for Statistical Computing. 2018a, R: a language and environment for statistical computing (https://www.R-project.org).

The R Foundation for Statistical Computing. 2018b, CRAN: The Comprehensive R Archive Network (https://cran.r-project.org/mirrors.html).

The R Foundation for Statistical Computing. 2018c, Rssa: a collection of methods for sungular spectrum analysis (http://cran.r-project.org/web/packages/Rssa).

Rodriguez, E., *et al.* 2004, *Mon. Not. Roy. Astron. Soc.*, **347**, 1317.

RStudio. 2018, RSTUDIO software (https://www.rstudio.com).

Sabin, L., and Zijlstra, A. A. 2006, *Mem. Soc. Astron. Ital.*, 77, 933.

Samus N. N., Kazarovets, E. V., Durlevich, O. V., Kireeva, N. N., and Pastukhova E. N. 2017, *Astron. Rep.*, **61**, 80, *General Catalogue of Variable Stars*: version GCVS 5.1 (http://www.sai.msu.su/gcvs/gcvs/index.htm).

Variable Star Observers League in Japan. 2018, VSOLJ variable star observation database (http://vsolj.cetus-net.org/database.html).

Various authors. 2010, *Statistics and its Interface*, **3** (No. 3).

Various authors. 2017, *Statistics and its Interface*, **10** (No. 1).

Vautard, M., and Ghil, M. 1989, *Phys. D: Nonlinear Phenomena*, **35**, 395.

Vautard, M., Yiou, P., and Ghil, M. 1992, *Phys. D: Nonlinear Phenomena*, **58**, 95.

Wenger, M., *et al.* 2000, *Astron. Astrophys., Suppl. Ser.*, **143**, 9.

Zhigljavsky, A. 2010, *Statistics and its Interface*, **3**, 255.

Appendix A: code examples

A.1. SVD code

Notes:

1. We recommend the use of "RSTUDIO" (2018) which provides a simple and highly efficient way of handling R code and results.

2. The user needs to set the path according to where the R system has been installed—see the code comment below.

3. The packages "tseriesChaos" and "Rssa" need to be installed from the "install" tab under "packages" in RSTUDIO.

4. The code should be saved as "XXXX.R" in the "User Defined Function" subdirectory of R when "XXXX" is a user chosen name.

5. Steps 1–6 are present to show what is going on behind the scenes in Step 7—in practical use only Step 7 is needed.

6. Comments are in italics, code in bold.

```
# Code: Basic SSA - matrix decomposition and grouping
rm(list=ls(all=TRUE))

# DEFINE YOUR PATH HERE
setwd("C:/Users/Geoff/Documents/R/data")
# END DEFINE YOUR PATH HERE

# User-defined function for averaging of minor diagonals—from Huffaker et
al. (2017) code 6.6
diag.ave<-function(mat, rowCount, colCount) {
   hold<-matrix(0,(rowCount+(colCount-1)))
   for(i in 1:(rowCount+(colCount-1))) {
   if(i==1) {d<-mat[1,1]}

   if(i>1 & i<=colCount) {d<-diag(mat[i:1,1:i])}

   if(i>colCount & i<=rowCount)  {d<-diag(mat[i:(i-(colCount-
1)),1:colCount])}

   if(i>rowCount & i<(rowCount+(colCount-1))) {
   d<-diag(mat[rowCount:(i-(colCount-1)),(i-(rowCount-1)):colCount])}

   if(i==(rowCount+(colCount-1))) {d<-mat[rowCount,colCount]}

   d.ave<-mean(d) #average minor diagonals
   hold[i,]<-d.ave
   } #end loop
   return(hold)
} #end function

# START START START START START START START START START START
START START
# USER INPUT USER INPUT USER INPUT USER INPUT USER INPUT
USER INPUT USER

# Read in data
ts<-read.csv("RZ Cas O minus C.csv")
x<-ts$OmCadj #x has ndata rows and 1 col

# dimension (number of columns) of the trajectory matrix
L = 200
# choose the number of eigenvectors (and reconstructed series) required
outputVecCount = 20

# end USER INPUT USER INPUT USER INPUT USER INPUT USER INPUT
USER INPUT

# step 1: construct trajectory matrix

library(tseriesChaos)
TM = embedd(x,L,1) #1=delay
```

```
ndata=length(TM[,1])

# step 2: lagged covariance matrix

lagCM = t(TM) %*% TM

# step 3: eigensystem of lagCM

eigensys = eigen(lagCM,symmetric=TRUE)
eigenvals = eigensys$values
eigenvecs = eigensys$vectors
eigenSet = cbind(eigenvals,eigenvecs)

orderedSet = order(eigenSet[,1],decreasing=TRUE)

ES = eigenSet[order(eigenSet[,1],decreasing=TRUE),] #sort in order of
eigenvalues

# calculate relative strength of EVs

sumLambdas = sum(eigenvals)
relativeEV = matrix(0,nrow=outputVecCount,ncol=1)
for (i in 1:outputVecCount) {relativeEV[i] = abs(ES[i,1])/sumLambdas}

write(relativeEV[1:outputVecCount], file = "BasicSSAdata.csv",
   ncolumns = outputVecCount,append = FALSE, sep = ",")

# PLOT: relative eigenvalue plots

plot(relativeEV[1:outputVecCount],log="y",type="b",col="black",lwd=2)

# calculate left eigenvectors of the trajectory matrix

left = matrix(0,nrow=ndata,ncol=outputVecCount)
for(i in 1:outputVecCount){
   left[,i] = TM %*% ES[,1+i]/sqrt(ES[i,1])
}

# step 4: now get the decomposition of the TM (trajectory matrices projected
# on important eigenvectors)

X = array(1:ndata*L*outputVecCount,dim=c(ndata,L,outputVecCount))
for(i in 1:outputVecCount){
 X[,,i] = sqrt(ES[i,1]) * left[,i] %*% t(ES[,1+i])
}

# step 5: reconstructed individual time-series (diagonal averaging)

actualNdata = ndata+L-1
recon = matrix(0,nrow=actualNdata,ncol=outputVecCount)
for (i in 1:outputVecCount) {
 recon[,i] = diag.ave(X[,,i],ndata,L)
}

# PLOT: plot of correlations

w<-cor(recon,y=NULL,use="everything",method="pearson")
library(corrplot)
corrplot(w,method="square")

# PLOT: miniplot of recon time series related to each EV

plotRow = round(sqrt(outputVecCount))
par(mfrow=c(plotRow,outputVecCount/plotRow))
for (i in 1:outputVecCount){
   plot(recon[,i],xlim=c(1,200),xlab="",
   ylab=paste("series ",toString(i),"; ",toString(round(1000*relativeEV
[i])/10),"%"),
   type="l",col="black",lwd=2) #plot 1st 20 time series for 200 periods
}
```

write time series output

```
write(t(recon), file = "BasicSSAdata.csv",#tmp
ncolumns = outputVecCount,append = TRUE, sep = ",")
```

A.2. Toeplitz code
Steps 1 and 2 in the above are replaced with the following:

```
#step 1: construct trajectory matrix
zero = seq(0,0,length.out=ndata) #used for padding
TM = matrix(0,nrow=ndata,ncol=L)
TM = cbind(x,append(x[2:ndata], zero[1:1], after=ndata-1))
for(j in 3:L){
    TM = cbind(TM,append(x[j:ndata],zero[1:j-1],after=ndata-j+1))
}
```

```
#step 2: lagged covariance matrix
lagCM = matrix(data=NA,nrow=L,ncol=L)
for(i in 1:L){
    for(j in 1:L){
    xsum = 0
    for (t in 1:(ndata-abs(i-j))){
    xsum = xsum + x[t]*x[(t+abs(i-j))]
    }
    xsum = xsum / (ndata-abs(i-j))
    lagCM[i,j] <- xsum
    }
}
```

A.3. Rssa code

```
#Code 6.9 from Huffaker et al. (2017), SSA: matrix decomposition and grouping
diagnostics
rm(list=ls(all=TRUE))
```

```
#Read in data
setwd("C:/Users/Geoff/Documents/R/data")
ts<-read.csv("RZ Cas O minus C.csv");
x<-ts$OmCadj
n = length(x)
```

```
#SSA Decomposition
#load Rssa R library from Install Packages
library(Rssa)
L=200
s<-ssa(x,L,kind="1d-ssa") #run Rssa 1d-ssa
#s<-ssa(x,L,kind="toeplitz-ssa") # alternatively run Rssa Toeplitz-ssa
```

```
#Run grouping diagnostics to group eigentriplets
#First visual diagnostic: Eigenspectrum
plot(s,numvalues=20,col="black",lwd=2) #plot 1st 20 largest
eigenvalues<-plot(s,numvalues=20,col="black",lwd=2)
```

```
#Second visual diagnostic: Eigenvector plots
plot(s,type="vectors",idx=1:20,xlim=c(1,200),col="black",lwd=2) #plot
1st 20 for 300 periods
```

```
#Weighted correlation matrix
plot(w<-wcor(s,groups=c(1:19))) #1st 20 eigentriplets
w.corr.res<-wcor(s,groups=c(1:20)) #table for 1st 10 eigentriplets
```

```
# write time series output
r.1<-reconstruct(s,groups=li
st(1,2,3,4,5,6,7,8,9,10,11,12,13,14,15,16,17,18,19,20))
recon.1<-r.1$F1
recon.2<-r.1$F2
recon.3<-r.1$F3
recon.4<-r.1$F4
recon.5<-r.1$F5
recon.6<-r.1$F6
recon.7<-r.1$F7
recon.8<-r.1$F8
recon.9<-r.1$F9
recon.10<-r.1$F10
recon.11<-r.1$F11
recon.12<-r.1$F12
recon.13<-r.1$F13
recon.14<-r.1$F14
recon.15<-r.1$F15
recon.16<-r.1$F16
recon.17<-r.1$F17
recon.18<-r.1$F18
recon.19<-r.1$F19
recon.20<-r.1$F20
```

```
tmp = vector("numeric",20)
write(c(1:20), file = "BasicRssadata.csv",ncolumns = 20,append = FALSE,
sep = ",")
for (i in 1:n) {
    tmp = c(recon.1[i],recon.2[i],recon.3[i],recon.4[i],recon.5[i],
    recon.6[i],recon.7[i],recon.8[i],recon.9[i],recon.10[i],
    recon.11[i],recon.12[i],recon.13[i],recon.14[i],recon.15[i],
    recon.16[i],recon.17[i],recon.18[i],recon.19[i],recon.20[i])
    write(t(tmp), file = "BasicRssadata.csv",ncolumns = 20,append = TRUE,
sep = ",")
}
```

Unmanned Aerial Systems for Variable Star Astronomical Observations

David H. Hinzel

Engineering Tecknowledgey Applications, LLC, 9315 Argent Court, Fairfax Station, VA 22039; daveeta1@cox.net; www.engtecknow.com

Received July 21, 2018; revised July 26, August 20, 2018; accepted September 24, 2018

Abstract   Variable star astronomy (and astronomy in general) has two problems: a low altitude problem and a high altitude problem. The low altitude problem concerns ground-based observatories. These observatories are limited by inclement weather, dust, wind, humidity, environmental and light pollution, and often times being in remote locations. Ideal locations are limited to dry and/or high elevation environments (e.g., the Atacama Desert in Chile). Locations such as low elevation, rainy, and polluted environments are undesirable for ground-based observatories. These problems can be resolved by spacecraft operating above the degrading effects of the atmosphere, but come at a very high price (the high altitude problem). Additionally, maintenance is impossible with space-based telescopes (e.g., Kepler with its degraded performance due to the loss of reaction wheel control). One potential solution to the low/high problems may be to utilize an Unmanned Aerial Vehicle (UAV) carrying a telescope payload. A modest sized UAV could easily carry a telescope system payload high above the ground environment at a fraction of the cost of spacecraft. This could permit essentially round-the-clock operation in virtually any location and in any type of environment. This will become increasingly important to both professional and amateur astronomers who will need quick access to telescopes for follow-up support of new astronomical observatories such as the Large Synoptic Survey Telescope (LSST) and the Transiting Exoplanet Survey Satellite (TESS) which will continuously generate enormous amounts of data.

1. Introduction

A potential solution to the low/high problem of ground-based and space-based astronomical observatories is to utilize a "moderate-altitude system" that operates above the degrading effects of the near-earth environment but at a fraction of the cost of space-based assets. Such a moderate-altitude system is the Unmanned Aerial System (UAS). Unmanned Aerial Systems consist of multiple components or segments: an Unmanned Aerial Vehicle (UAV) segment, a ground-based UAV control and status segment, a ground-based "mission product" data collection and processing segment, and a ground-based segment for UAV payload control and status. An Unmanned Aerial Vehicle, commonly known as a drone, is an aircraft without a human pilot aboard, but controlled by humans from the ground control stations. The UAV would host an astronomical telescope payload, not unlike what is used at ground observatories. Figure 1 shows the high level system architecture of an Unmanned Aerial System operated as an airborne telescope platform.

2. Background

Unmanned Aerial Vehicles (UAVs) have existed for many decades, but have achieved impressive technical advances in approximately the last 20 years. This is primarily the result of use of UAVs by the U.S. military since the terrorist attacks of September 11, 2001. These systems have been used for surveillance and intelligence gathering missions as well as delivery of weapons-on-target in Iraq, Afghanistan, Syria, and other war zones without endangering human pilots and aircraft crew members.

In addition to the military uses and applications, drones have become the vehicle of choice for many non-military, civilian, and commercial applications and have become a household word that most people are familiar with. As the drones continue to evolve

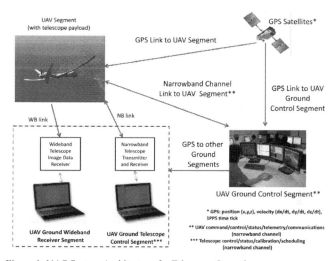

Figure 1. UAS System Architecture for Telescope Operations.

technologically, they are becoming capable of carrying more advanced payloads on smaller platforms at decreasing costs. What would have been difficult, if not impossible, even a few years ago is now rapidly becoming feasible at affordable costs.

One application for Unmanned Aerial Systems that appears not to have been seriously addressed to-date is their use as medium altitude astronomical observation platforms. Optical and electro-optical payloads are relatively common on drones for ground surveillance and eyes-on-target military and intelligence applications. However, astronomical observation applications with the optics "pointing the other way" seems to have not received much attention. Both types of optical payload systems share common problems and solutions. This paper will address many of the issues of importance, from an engineering perspective, necessary to utilize Unmanned Aerial Systems for astronomical observations.

As previously mentioned, Unmanned Aerial Systems consist of various component segments. While the exact number

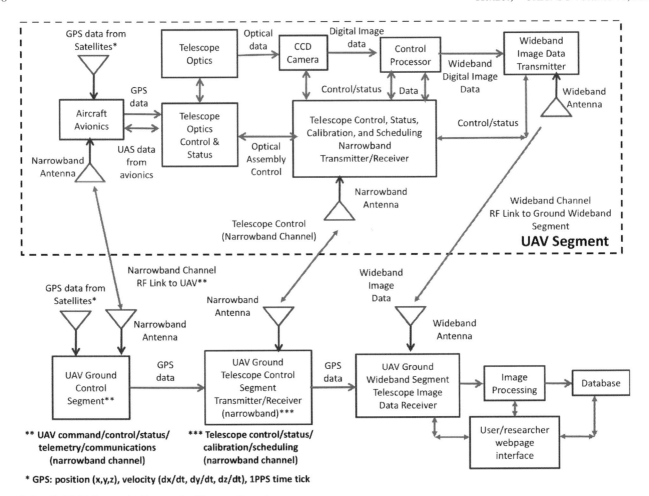

Figure 2. Detailed UAS System Architecture for Telescope Operations.

of segments depends upon the level of detail required, the overall system can be broken down into four major parts: an Unmanned Aerial Vehicle (UAV) segment, a ground-based UAV control and status segment, a ground-based mission product data collection and processing segment, and a ground-based segment for UAV payload control and status. Additionally, all segments rely upon GPS information, but this is not considered to be part of the UAS. This overview is shown in Figure 1. Below is the detailed description of each segment. However, most of the emphasis will be placed on the UAV segment since it is the astronomical observation platform of interest, with the other segments acting in supporting roles. The detailed UAS systems architecture for astronomical telescope operations is shown in Figure 2.

3. Instrumentation and methods

3.1. Unmanned aerial vehicle segment

The UAV airborne component or segment is the "housing" for the telescope payload in the same way that the ground-based observatory and spacecraft are the "housings" for ground-based and space-based telescopes, respectively. Although there are potentially numerous possible candidates for the airborne segment, one excellent example is the NASA Altair UAV (NASA 2015). The Altair, a high altitude version of the military Predator B, was specifically designed as an unmanned platform

for both scientific and commercial research missions that require endurance, reliability, and increased payload capacity. Figures 3 and 4 show the Altair UAV in flight and on the ground with its payload bay open, respectively. The Altair has an 86-foot wingspan, a 36-foot length, can fly up to 52,000 feet with airspeed of 210 knots, and has an airborne endurance of 32 hours. Additionally, it has a gross weight of 7,000 pounds and can carry a payload up to 750 pounds (sensors, communications, radar, and imaging/telescope equipment). It incorporates redundant fault-tolerant flight control and avionics systems for increased reliability, GPS and INS (Inertial Navigation System), an automated collision avoidance system, and air traffic control voice communications for flights in National Airspace. Finally, it can be remotely piloted or operated fully autonomously.

3.1.1. Telescope optical system

The UAV telescope optical system would not be simply mounting a ground observatory type telescope in the aircraft (i.e., installing an AAVSOnet type telescope). This payload would necessarily incorporate adaptive and/or active optics. Adaptive optics is a technology used to improve the performance of optical systems by reducing the effect of incoming wavefront distortion by deforming a mirror in order to compensate for the distortion. Active optics is a technology used with reflecting telescopes which actively shapes the telescope mirrors to prevent deformation due to external influences such as wind,

Figure 3. The NASA Altair UAV in flight.

Figure 4. The NASA Altair UAV payload bay.

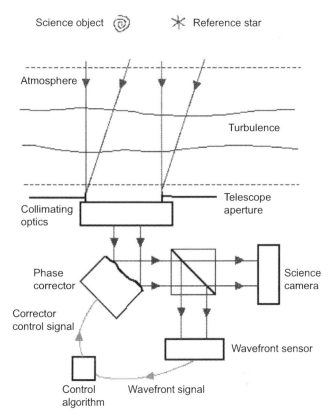

Figure 5. Adaptive optics architecture.

temperature, vibration, and mechanical stress. If the UAV segment is always flown at its maximum altitude of 52,000 feet, adaptive optics may not be needed since at that altitude it would be above all but a few percent of the earth's atmosphere (60,000 feet is above all but 1% of the earth's atmosphere). However, if flown at lower altitudes, adaptive optics may be necessary. Therefore, in order to account for all reasonable operational scenarios, it will be assumed that both adaptive and active optics are required (Wikipedia 2018).

Adaptive optics works by measuring the distortions in a wavefront and compensating for them with a device that corrects those errors such as a liquid crystal array or a deformable mirror. Deformable mirrors utilizing micro-electro-mechanical systems (MEMS) are currently the most widely used technology in wavefront shaping applications due to the versatile, high resolution wavefront correction that they afford. Active optics utilizes an array of actuators attached to the rear side of the mirror and applies variable forces to the mirror body to keep reflecting surfaces in the correct shape. The system keeps a mirror in its optimal shape against environmental forces such as wind, sag, thermal expansion, vibration, acceleration and gravitational stresses, and telescope axis deformation. These are all significant concerns in the UAV aircraft operational environment. Active optics compensate for these distorting forces that change relatively slowly, on the time scale of seconds, thereby keeping the mirror actively still in its optimal shape. Adaptive optics operates on shorter time scales (milliseconds) to compensate for atmospheric effects, rather than for mirror

deformations. Adaptive optics and active optics can both be incorporated in the telescope optical path, since the former uses smaller corrective mirrors (secondary mirrors) while the latter is generally applied to the reflector primary mirror. Using both, atmospheric wavefront and aircraft environmental effects can be compensated for and corrected. Figure 5 illustrates generic adaptive optics architecture.

One of the most important issues for high altitude astronomical observations will be the use of guide stars as an optical reference for science imaging. Due to the continuous motion of the UAV, guide stars may be even more important than for ground-based observations. Since objects in science imaging may be too faint to be used as a reference for measuring the shape of optical wavefronts, bright guide stars in close proximity to the target stars need to be used. Since both the target and guide stars pass through the same atmospheric turbulence, the guide star can be used as a calibration reference source for the target star, thereby applying small corrections to the adaptive optics system. Additionally, since the UAV will necessarily use an optical window as the telescope "radome," the bright reference guide stars will be used to "calibrate out" the degrading effects of the optical window material (scratches, material imperfections, optical aberrations such as reflections, and so on).

In the situation where there are no useable natural guide stars close to the target star, artificial guide stars can be generated by using an onboard laser beam as the reference light source. Laser guide stars work by exciting atoms in the upper atmosphere, which then produce optical backscatter that can be detected by the onboard adaptive optics. The laser guide stars can then be used as a wavefront reference in the same way as a natural guide star. The weaker natural reference stars are still required for image position information (plate solving/pattern matching).

Two examples of optics payloads flown on high altitude aircraft are given in Appendix C.

3.1.2. Science image and data processing

The output of the adaptive/active optical system is optical data similar to most other telescopes, but presumably corrected for wavefront and UAV environmental degradations. As with other telescope systems, the optical data are immediately captured on a CCD camera, including any UBVRI (or other, Ha, for example) filtering that is required by the telescope user. The CCD camera output is digital image data which are then sent to the UAV telescope control processor. The control processor is the heart and brain of the telescope system since it coordinates all of the astronomical digital image data with the telescope control protocols being uplinked from the UAV telescope ground station.

As shown in Figure 2, the telescope control data, received from the ground, command the optics, receive optics performance status, perform optics calibrations (darks, flats, etc.), and schedule user activity (number/type/duration of images required, filter combinations, etc.). Additionally, the telescope control provides a feedback mechanism to the telescope optics control and status system (the UAV equivalent of a computerized telescope equatorial mount). This is also combined with the aircraft avionics data and the received UAV GPS data. The combination of these various inputs to the telescope optics keeps the optics stabilized against aircraft motion, vibration, shock, and acceleration as well as providing GPS data for "locating" the telescope as it flies its mission. The GPS provides x, y, and z position, dx/dt, dy/dt and dz/dt velocity, and a 1 PPS (pulse per second) time tick to synchronize all functions within the UAV telescope system as well as the overall UAV avionics system. All of the UAV telescope payload functions are controlled from the UAV ground telescope control segment via a narrowband channel RF (radio frequency) transmitter/receiver on the UAV segment.

Finally, the wideband digital image data (science images) from the telescope system are downlinked via an onboard wideband image data transmitter to the UAV ground wideband segment for further image processing as may be required. No wideband receiver is needed here since no wideband image data is uplinked to the UAV.

3.2. Ground-based UAV control and status segment

The UAV Ground Control Segment is used to control the UAV airborne segment. There are few, if any, direct interfaces with the telescope payload. The UAV Ground Control Segment acts as the aircraft "pilot" and air traffic control tower, similar to that for commercial or military aircraft operations. Instead of onboard human pilots and aircrew, all UAV flight operations are orchestrated remotely from the ground. The UAV has, at least in the case of the NASA Altair vehicle, redundant fault-tolerant flight control and avionics systems for increased reliability, GPS and INS (Inertial Navigation System), an automated collision avoidance system, and air traffic control voice communications for flights in National Airspace. It can be remotely piloted or operated fully autonomously. The Ground Control Segment interfaces with the UAV for command, control, flight status, operational telemetry, and voice/data communications. As a result, the UAV would appear as a normal commercial aircraft to national and international air traffic control systems.

3.3. Ground-based segment for UAV telescope payload control and status

Unlike the UAV Ground Control Segment which has little to do with telescope payload operation, the UAV Ground Telescope Control Segment handles all of the onboard telescope operations. These functions are similar to a ground based robotic telescope or perhaps satellite based telescope (but most likely considerably simpler then space assets). This UAS segment uses a narrowband communications channel to uplink telescope commands. This system performs telescope command, control, performance status, calibration, scheduling, and miscellaneous telescope payload overhead functions. This communicates directly with the UAV telescope control system on the aircraft as described above. The combination of the ground-based and aircraft-based telescope control system can be thought of as a distributed computerized equatorial mount for the airborne telescope. The UAV Ground Telescope Control Segment also receives GPS data, thereby keeping this segment synchronized with all the other UAS segments. Just as pilots "think" they are in the cockpit flying the aircraft, telescope users "think" they are operating a ground-based robotic telescope.

3.4. Ground-based wideband data collection and processing segment

The fourth UAS segment is the UAV Ground-based Wideband Data Collection and Processing Segment. This function again has nothing to do with the UAV aircraft operation, but is necessary for wideband image (science image) ground processing and post-processing. This segment utilizes a wideband RF receiver to acquire and process science data. Again, this segment receives GPS data to stay synchronized with the other segments, although in all likelihood only the 1 PPS timing information will be necessary. No RF transmitter is necessary in this segment since no wideband image data are sent back to the UAV. The processed/post-processed science image data are finally sent to astronomical databases for use by researchers, similar to the wide variety of astronomical databases currently in existence (e.g., Kepler, ASAS, ASAS-SN, NSVS, etc.).

4. Concept of operations and results

Based upon the UAS System architecture for telescope operations as presented above and shown in detail in Figure 2, a Concept of Operation (CONOPS) can be developed incorporating both the UAV flight parameters and telescope observation techniques. The CONOPS will include 1) UAV flight procedures necessary for telescope operation at altitude, 2) orbital scenarios, 3) optics calibration, and 4) telescope operation by the user/researcher. It will be assumed that the NASA Altair vehicle is used for this mission and flight procedures discussed below will reflect those of the Altair.

4.1. UAV flight procedures necessary for telescope operations at altitude

The UAV airborne segment of the overall UAS system, which carries the telescope payload to its operational altitude of 52,000 feet, can take off from any runway available as long as the UAV Ground Control Segment is in close proximity. The UAV and UAV Ground Control Segment do not necessarily have to be close to the UAV Ground Telescope Control Segment and/ or the UAV Ground Wideband Segment telescope image data receiver, although there would be some advantages in doing so. The telescope payload would be dormant, powered down, and physically secured during all UAV takeoff (and landing) flight operations. As the UAV is climbing to altitude, the telescope payload bay would begin environmental stabilization procedures. These would include temperature and humidity control, mechanical vibration and shock damping, and condensation management (primarily when landing where extreme condensation can be a serious problem, even producing "rain" within the payload bay). Appendix A outlines mitigation procedures for controlling condensation.

4.2. Orbital scenarios

Upon reaching the desired altitude and after all equipment and the payload environment have stabilized, the UAV would be positioned into its operational "orbit." High altitude UAVs are typically flown in a long-duration loitering orbit around ground-based targets. Similar orbits would be used for astronomical observations. These operational orbits are typically long elliptical paths or four-sided "box" paths. The idea is to maintain the flight profile as straight as possible for as long as possible to minimize turning or banking maneuvers. With a 32-hour airborne endurance time, extremely long, stable flight paths should be possible.

After insertion into a stable operational orbit, the telescope payload will become operational. All telescope systems can be powered up, including all of the RF transmitters and receivers required to transmit and receive telescope commands to and from the ground segment as well as the wideband image transmitter for downlinking science images. At this point, telescope optics calibration can begin.

4.3. Optics calibration

Optics image calibration should be done when the UAV has reached operational altitude and the temperature and humidity inside the telescope payload bay have stabilized. In particular the CCD camera cooler should be allowed to stabilize prior to taking the image frames. Approximately 0.5 hour is the recommended time for payload stabilization. The optics calibration is similar to that performed with ground-based systems, i.e. bias frames, dark frames, and flat frames. Appendix B outlines calibration procedures based upon that recommended by the AAVSO. After calibration the telescope is ready for user/researcher operation.

4.4. Telescope operation by the user/researcher

The user/researcher would use the airborne telescope in a manner similar to that of a ground-based robotic telescope system such as AAVSOnet or iTelescope. Specifically, the users/ researchers would access the telescope through a webpage that allows them to make reservations for time in the future, schedule various observing scenarios such as multiple short (time domain) images or long exposure images, UBVRI or other filter selection for imaging, manual optics calibration if desired, plate solving/pattern matching, focusing, and image download. Image processing by the user would then be done with appropriate personal software or access to analysis software such as VPHOT.

At the conclusion of an imaging session at altitude, the procedure for landing the UAV is essentially the reverse of the takeoff procedure. The telescope payload will be physically and mechanically secured, and powered down. Additionally, condensation control and mitigation will be started in order to protect the equipment from the adverse effects of condensation as the UAV decreases altitude. This is discussed in Appendix A.

5. Discussion and future directions

The implementation of moderate-to-high altitude UAVs carrying telescope payloads for astronomical observations appears not to have been seriously addressed to date as an application of rapidly advancing drone technology. This application allows observations to be performed above most, if not all, degrading atmospheric effects at a fraction of the cost of an equivalent space-based asset. Such an application can be realized with currently exiting scientific and technological resources at moderate cost. Full implementation of a system similar to that discussed above will require no new principles of physics nor advanced technologies that currently do not exist, i.e., needs nothing new to be discovered or invented. While new technologies would undoubtedly be helpful, nothing new is necessary.

Therefore, the current and future challenge will not be scientific or technological, but rather one of focused attention on solving the problem with adequate funding. This is almost always the hardest part of bringing new ideas to fruition. By combining private sector innovation with academic research capabilities and adequate capital resources from committed investors, Unmanned Aerial Systems for astronomical observation could become a reality in the near future.

6. Conclusions

One potential solution to the low/high problems may be to utilize an Unmanned Aerial Vehicle (UAV) carrying a telescope payload. A modest sized UAV could easily carry a telescope system payload high above the ground environment at a fraction of the cost of spacecraft. This could permit essentially round-the-clock operation in virtually any location and in any type of environment. This will become increasingly important to both professional and amateur astronomers who will need quick access to telescopes for follow-up support of new astronomical observatories such as the Large Synoptic Survey Telescope (LSST) and the Transiting Exoplanet Survey Satellite (TESS) which will continuously generate enormous amounts of data. Quick access to telescope systems that are not affected by environmental factors and enjoy very high duty cycles will become extremely valuable. Examples of optics payloads on two other high altitude aircraft are discussed in Appendix C.

7. Acknowledgements

The author would like to thank the U.S. Department of Defense (DoD) Small Business Innovative Research (SBIR) program for the award of contract number N00039-03-C-0010, Anti-Terrorism Technologies for Asymmetric Naval Warfare: Detection, Indication, and Warning, managed by the Space and Naval Warfare Systems Command (SPAWAR). This project demonstrated that advanced operational airborne systems and capabilities can be quickly designed, developed, and flight tested with existing technologies at reasonable costs. Specifically, this project successfully implemented an advanced electronic signal intelligence collection/geolocation system payload on a high altitude Unmanned Aerial Vehicle. This system included the UAV and all ground control and data collection/processing assets.

Additional thanks go to the author's coauthors of U.S. Patent Number 6559530 (Hinzel *et al.* 2003) who were involved in the development of Micro-Electro-Mechanical Systems (MEMS) technology. MEMS technology is critical for adaptive optics systems. Finally, the author wishes to thank the AAVSO for the publication of the *AAVSO Guide to CCD Photometry* (AAVSO 2014) which provides the complete procedure for accurate CCD photometry, including all of the calibration procedures.

References

AAVSO. 2014, *AAVSO Guide to CCD Photometry* (version 1.1; https://www.aavso.org/ccd-photometry-gude).

Hinzel, D., Goldsmith, C., and Linder, L. 2003, "Methods of Integrating MEMS Devices with Low-Resistivity Silicon Substrates", Patent # 6559530 (May 2003).

NASA. 2015, NASA Armstrong Fact Sheet: Altair (https://www.nasa.gov/centers/armstrong/news/FactSheets/FS-073-DFRC.html).

Wikipedia. 2018, Active-optics, Adaptive-optics (https://en.wikipedia.org/wiki/Active-optics; https://en.wikipedia.org/wiki/Adaptive-optics).

Appendix A: Condensation mitigation

For electro-optics systems, any residual moisture within the internal cavity or enclosure operated in the field and/or at altitude could produce disruptive condensation that fogs mirrors and lenses, which effectively could blind the equipment in critical situations. The other concern with condensation is corrosion, which is just as destructive because it can degrade performance and shorten system lifespan. Often used in commercial and military applications, electro-optics systems are mounted on aircraft, helicopters, missiles, or transported at high elevations where extremely low temperatures and air pressure can cause condensation even with minimal moisture present. With so much at stake, manufacturers of laser, imaging, camera, and other optical systems are increasingly mandating a nitrogen purge to wring the moisture out of enclosures and cavities

before these systems are deployed to the field. However, this problem is still potentially serious when systems are operated at high altitude, even if measures have been taken during the manufacturing process to minimize the condensation problem. In a nitrogen purge, ultra dry nitrogen with a dew point of –70 degrees Celsius is introduced under pressure into an enclosure or cavity to remove moisture and create a much drier internal environment than standard desiccant can achieve. Nitrogen purging is accomplished through commercially available purging systems or ad hoc systems created by the engineers designing the product itself. The concept of a nitrogen purge is essentially to "squeeze" the internal components like a sponge to remove any residual humidity or moisture out of the system and then seal it up to keep the internal cavity moisture-free during its operational life.

It is a common misconception that the majority of the moisture in a sealed cavity or enclosure is contained in the empty volume of air. In fact, the majority of the moisture is contained in the hygroscopic materials, such as common internal plastic circuit boards or other plastic components within the enclosure. Hygroscopic plastics readily absorb moisture from the atmosphere and can release that moisture under temperature cycling and other environmental factors.

The internal electronics are the main culprit for much of residual moisture and must be remedied with a nitrogen purge. A nitrogen purge enters the cavity or enclosure through a single port and is pressurized to a pre-determined level before a valve opens and the gas flows back into the unit. There it passes a dew point monitor and displays the current dew-point temperature. The nitrogen is then vented to the atmosphere and a new cycle commences. This cycling continues until the equipment reaches the required dew-point level, at which point it automatically shuts off.

Appendix B: Optics calibration

Bias Frames: Bias frames should be done in a dark environment with the shutter closed. Exposure should be zero seconds or as short as possible. Approximately 100 images should be taken and averaged together to create a Master Bias.

Dark Frames: Dark frames should be done in a dark environment with the shutter closed, with exposure time as long or longer than the science images. Twenty or more images should be taken. If combining into a raw Master Dark use this only with science frames of the same exposure and do not use the Master Bias. If combining into a Master Dark, subtract the Master Bias from each, then average- or median-combine them all to create a Master Dark for use with science frames of equal or shorter exposure. Use this with the Master Bias in calibration.

Flat Frames: Flat frames should be taken with a uniform, calibrated light source within the telescope payload bay. The focus should be the same as that of the science images and exposure time should result in about half of the full well depth of the CCD. Ten or more images should be taken for each filter and averaged- or median-combined together. Subtract a Master Dark and Master Bias to create a Master Flat.

Appendix C: Optics payloads on high altitude aircraft

The application of UAS technologies and systems to variable star astronomical observations would not be the first time that optics payloads have been deployed on high altitude aircraft. Two notable examples are the military/homeland security systems which utilize what are termed Multispectral Targeting Systems (MTS) and the SOFIA system developed by NASA and the German Aerospace Center (DLR). SOFIA is the Stratospheric Observatory for Infrared Astronomy.

While the details of the military/homeland security MTS payloads are classified and not publicly available, some indication of this capability may be inferred from what is known about certain recent anti-terror operations utilizing the Predator B, Predator XP, Grey Eagle, and other high altitude systems. Figure 6 shows a typical MTS mounted on the underside of the Predator vehicle.

This system utilizes a 12-inch sensor turret weighing less than 60 pounds that sees in infrared and the visible spectrum and delivers intelligence in high-definition, full motion video. Its camera contains 4096 × 4096 pixels with a field of view of

Figure 6. Predator Multispectral Targeting System.

11.4 × 11.4 degrees. Other MTS units have selectable fields of view, for example a wide FOV of 27.7 degrees, a medium FOV of 6.0 degrees, and a narrow FOV of 1.02 degrees. If it assumed that the Predator is operating at its maximum altitude of 52,000 feet and can clearly detect and track in real-time a human being who is approximately 6 feet tall and running at high speed, this will give an indication of the resolution that is operationally achievable. Undoubtedly, the actual classified performance would be significantly better than that.

SOFIA is a multi-sensor platform flown in a modified Boeing 747 aircraft carrying a 2.7-meter (106-inch) reflecting telescope. The telescope consists of a parabolic primary mirror and a hyperbolic secondary mirror in a bent Cassegrain configuration, with two foci (the nominal IR focus and an additional visible light focus for guiding). The IR image is fed into a focal plane imager which is a 1024 × 1024 pixel science grade CCD sensor. It covers the 360–1100 nm wavelength range, has a plate scale of 0.51 arcsec/pixel, and a square field of view of 8.7 × 8.7 arcminutes. Five Sloan Digital Sky Survey filters—u', g', r', i', z'—and a Schott RG1000 NIR cut-on filter are available. The system f-ratio is 19.6 and the primary mirror f-ratio is 1.28. Telescope elevation range is approximately 23–57 degrees with a field of view of 8 arcminutes.

In addition to the telescope itself, SOFIA carries several science instruments, including a mid-IR Echelle spectrometer, a far-IR grating spectrometer, a mid-IR camera and grism spectrometer, a far-IR heterodyne spectrometer, a far-IR bolometer camera and polarimeter, and a mid-IR bolometer spectrometer.

This document does not contain technology or technical data controlled under either the U.S. International Traffic in Arms Regulations (ITARs) or the U.S. Export Administration Regulations E17-76YV.

New Variables Discovered by Data Mining Images Taken During Recent Asteroid Photometric Surveys at the Astronomical Observatory of the University of Siena: Results for the Year 2017

Alessandro Marchini
Astronomical Observatory, DSFTA, University of Siena (K54), Siena, Italy; alessandro.marchini@unisi.it

Riccardo Papini
Wild Boar Remote Observatory (K49), San Casciano in val di Pesa, Florence, Italy

Fabio Salvaggio
Wild Boar Remote Observatory (K49), Saronno, Italy; Gruppo Astrofili Catanesi, Catania, Italy

Claudio Arena
Private Remote Observatory, Catania, Italy; Gruppo Astrofili Catanesi, Catania, Italy

Davide Agnetti
Osservatorio Aldo Agnetti di Lomazzo, Como, Italy

Mauro Bachini
Giacomo Succi
Osservatorio Astronomico di Tavolaia (A29), Santa Maria a Monte, Pisa, Italy

Massimo Banfi
Osservatorio di Nova Milanese (A25), Nova Milanese, Italy; Osservatorio delle Prealpi orobiche (A36), Ganda di Aviatico, Italy

Received May 1, 2018; revised June 6, 2018; accepted June 16, 2018

Abstract   This paper continues the publication of the list of the new variables discovered at Astronomical Observatory, DSFTA, University of Siena, while observing asteroids for determining their rotational periods. Further observations of these new variables are strongly encouraged in order to better characterize these stars, especially those showing non-ordinary light curves.

1. Introduction

The most essential activity at the Astronomical Observatory of the University of Siena, within the facilities of the Department of Physical Sciences, Earth, and Environment (DSFTA 2018), is mentoring the students in astronomy lab activities. Every month students attend CCD observing sessions with of the purpose of getting time-series photometry of asteroids, exoplanets, and variables. The large number of CCD images collected this way also enabled us to plot light curves of all the variable stars detectable in the images and check for new variables. If any was found, the variable was added to the AAVSO International Variable Star Index (VSX; Watson *et al.* 2014), to share them with the larger community of professional and amateur astronomers.

2. Instrumentation and methods

All the variables were discovered in the images taken at the Astronomical Observatory of the University of Siena using a Clear filter that transmits all wavelengths from UV to IR, since the main goal of the observations was the photometric study of faint asteroids to determine their synodic rotational period. As discussed in our previous paper (Papini *et al.* 2015), where the reader can find a detailed description of the strategy which characterizes our observations, once a new variable was found, aperture photometry was performed on each subset of data. Magnitudes are given as CV, which designates observations made without filter or using a Clear filter, but using V magnitudes for the comparison stars from available catalogues. In such a way the result will be closer to V but will vary depending on the sensitivity of the observer's setup and the color of the comparison stars.

For this reason, we merged our data with those available on-line from the main surveys. The most useful surveys turned out to be ASAS-3 (All Sky Automated Survey; Pojmański 2002), CRTS (Catalina Real-Time Transient Survey; Drake 2014), and NSVS (Northern Sky Variability Survey; Wozniak 2004). A special mention is made of the GAIA survey (Gaia Collaboration *et al.* 2016), whose Data Release 2 (Gaia Collaboration *et al.* 2018; Lindegren *et al.* 2018) arrived while this article was being prepared. GAIA DR2 has permitted including more information about the new variable stars presented in this work, such as their distances, as reported in Table 2.

Since photometric filters used in these surveys were different, it was mandatory to set a constant zero-point to fit

Table 1. Observers and main features of the instruments used.

Observer	Telescope*	CCD
Agnetti	28cm SCT f/10	Sbig ST 10
Arena	20cm NEW f/5	Atik 314L+
Bachini, Succi (A29)	40cm NEW f/5	DTA Discovery+ 260
Banfi (A25)	25cm SCT f/5	Sbig ST-7
Banfi (A36)	50cm NEW f/5	Sbig ST-9
Marchini (K54)	30cm MCT f/5.6	Sbig STL-6303E

* MCT = Maksutov-Cassegrain, NEW = Newton, SCT = Schmidt-Cassegrain

all the available data. The main elements presented in this work are independent of absolute magnitude, and therefore we decided to shift our data vertically, adding the difference between the average of the survey magnitudes and the average of the differential magnitudes worked out from our images. However, when the light curve phased against the period was not complete, we asked members of the Variable Star Section of the Unione Astrofili Italiani (SSV-UAI 2018) to follow up on the variables and collect data for the "missing" part of the light curve. Given the faint magnitude of the variable, we accepted unfiltered observations and shifted as described above. Each observer performed his own photometric analysis using the same reference stars (generally 3–4). Table 1 lists the observers' names and the main features of their instruments.

3. Recent discovery list and results

In the accompanying list (Table 2), we present the 24 new variables discovered during 2017, which, added to the previously discussed variables in our papers (Papini *et al.* 2015,

2017), bring the total to 95 variables discovered since 2015. For the statistics, of the 24 variables, 16 are eclipsing binaries (one of EA type, 11 EW, 4 EB) and 8 are short period pulsators (one of RRab type, 4 DSCT, 3 HADS).

In the following sections, we discuss briefly the only star with peculiar behavior, and present the light curves of the most representative type of variables.

3.1. UCAC4 557-036373

UCAC4 557-036373 is an EW binary system with a period of about 0.39344 day that has a low amplitude light curve variation between magnitude 15.43 and 15.69 CV. It shows clearly the O'Connell effect (O'Connell 1951; Liu and Yang 2003) with the two maxima at different amplitudes. Data from surveys were not available for this star. Figure 1 shows the light curve phased with the main period of the binary.

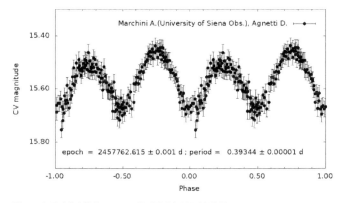

Figure 1. Folded light curve of UCAC4 557-036373.

Table 2. Main information and results for the new variables discovered.

Star (VSX identifier)	R.A. (J2000) h m s	Dec. (J2000) ° ' "	Const.	Parallax (mas)	CV Mag	Period (days)	Epoch (HJD–2450000)	Type
UCAC4 555-035787	06 59 09.13	+20 56 51.2	Gem	0.6879 ± 0.0698	15.16–15.70	0.38623 ± 0.00004	7762.6100 ± 0.0002	EW
UCAC4 557-036373	07 01 12.92	+21 17 32.7	Gem	0.4059 ± 0.0633	15.43–15.69	0.39344 ± 0.00001	7762.6150 ± 0.0008	EW
UCAC4 555-036219	07 01 38.40	+20 48 25.0	Gem	0.7917 ± 0.1300	15.97–16.30	0.37451 ± 0.00003	7760.6260 ± 0.0003	EW
GSC 01356-00372	07 02 31.80	+20 48 30.8	Gem	0.5267 ± 0.0392	13.41–13.51	0.081182 ± 0.000004	7759.3714 ± 0.0002	DSCT
GSC 01957-00131	09 17 33.89	+27 41 53.6	Cnc	0.0784 ± 0.0987	13.86–14.09	0.5060 ± 0.0001	7799.3868 ± 0.0002	EB
GSC 05536-00897	13 05 19.16	–09 09 18.9	Vir	0.5310 ± 0.0393	13.92–13.98	0.04562 ± 0.00006	7861.4271 ± 0.0004	DSCT
CMC15 J145002.3-051256	14 50 02.40	–05 12 56.0	Lib	0.4993 ± 0.0833	16.35–16.82	0.366271 ± 0.000004	7865.5445 ± 0.0003	EB
UCAC4 441-061555	15 50 44.36	–01 56 22.5	Ser		15.27–15.58	0.234492 ± 0.000002	7873.5057 ± 0.0004	EW
GSC 05627-00080	16 28 56.49	–08 07 27.1	Oph	0.4737 ± 0.3780	13.60–13.98	0.315999 ± 0.000005	7895.4496 ± 0.0003	EW
GSC 05627-00248	16 29 48.01	–07 45 11.4	Oph	0.5621 ± 0.0315	13.85–14.15	0.525977 ± 0.000004	7912.4069 ± 0.0003	EB
CMC15 J163041.4-080658	16 30 41.49	–08 06 58.9	Oph	0.0573 ± 0.1123	16.22–16.78	0.062443 ± 0.000004	7895.4461 ± 0.0005	HADS
UCAC4 410-066217	16 32 23.19	–08 01 43.3	Oph	1.1583 ± 0.0545	15.03–15.47	0.315450 ± 0.000004	7900.4089 ± 0.0003	EW
UCAC4 460-061118	16 51 31.20	+01 53 25.7	Oph	0.1344 ± 0.0834	16.25–16.60	0.066938 ± 0.000001	7899.5595 ± 0.0004	HADS
CMC15 J172111.9-045046	17 21 11.95	–04 50 46.1	Oph	0.1530 ± 0.1083	15.97–16.45	0.111612 ± 0.000001	7889.4210 ± 0.0004	HADS
UCAC4 428-070068	17 22 31.18	–04 32 53.5	Oph	0.2994 ± 0.0565	14.43–14.74	0.624796 ± 0.000005	7891.5091 ± 0.0004	RRAB
CMC15 J172246.1-043401	17 22 46.20	–04 34 01.1	Oph	0.8155 ± 0.1243	15.83–16.45	0.315587 ± 0.000006	7889.5333 ± 0.0004	EW
UCAC4 370-097050	17 38 28.90	–16 09 01.8	Oph	1.3581 ± 0.0369	14.30–14.80	0.358423 ± 0.000003	7924.4495 ± 0.0002	EW
UCAC4 369-097914	17 39 11.15	–16 16 25.2	Oph	0.7032 ± 0.0263	13.60–14.17	0.870247 ± 0.000006	7922.4514 ± 0.0005	EB
GSC 05117-01301	18 39 47.51	–02 45 05.8	Ser	2.4761 ± 0.0249	13.85–14.55	0.547939 ± 0.000002	7935.3423 ± 0.0001	EA
GSC 05117-00326	18 40 45.36	–02 26 19.5	Ser	0.4171 ± 0.0251	14.41–14.52	0.119096 ± 0.000004	7930.4890 ± 0.0003	DSCT
UCAC4 641-065317	19 06 44.58	+38 10 12.9	Lyr	0.6471 ± 0.0184	13.92–14.61	0.503517 ± 0.000003	7906.5021 ± 0.0002	EW
CMC15 J190719.6+375515	19 07 19.60	+37 55 15.5	Lyr	0.5615 ± 0.0451	16.37–16.92	0.285236 ± 0.000003	7907.4353 ± 0.0002	EW
UCAC4 641-065553	19 08 00.32	+38 01 57.1	Lyr	0.3778 ± 0.0343	15.69–16.24	0.398495 ± 0.000004	7907.5396 ± 0.0003	EW
UCAC4 409-132318	20 38 50.28	–08 22 42.1	Aqr	0.2866 ± 0.0420	15.09–15.20	0.058415 ± 0.000004	7951.4911 ± 0.0003	DSCT

Note: The column "Parallax" is derived from Gaia Data Release 2 data, recently available, and the value is expressed in milli-arcseconds. The column CV Mag is the magnitude range expressed in Clear (unfiltered) band aligned at V band, as explained in Section 2.

3.2. Eclipsing binaries

Since there are no stars in this class that show peculiar features or behavior, we will discuss in this section a few typical stars for each main subtype. GSC 05117-01301 is an eclipsing binary of EA type with a period of about 0.547939 day and a large amplitude light curve variation between magnitude 13.85 and 14.55 CV. Minima are quite similar in depth. No survey data were available for this star. Figure 2 shows the light curve phased with the main period of the binary.

GSC 05627-000248 is an eclipsing binary of EB type with a period of about 0.525977 day and an amplitude light curve variation between magnitude 13.85 and 14.15 CV. Minima are quite different in depth. Survey data from CRTS were available for this star and were added to our data. Figure 3 shows the light curve phased with the main period of the binary.

UCAC4 370-097050 is an eclipsing binary of EW type with a period of about 0.358423 day and a large amplitude light curve variation between magnitude 14.30 and 14.80 CV. Minima are slightly different in depth. No survey data were available for this star. Figure 4 shows the light curve phased with the main period of the binary.

3.3. Short period pulsators

As with the eclipsing binaries, there are no stars in this class that show peculiar features or behavior, and therefore we will discuss in this section a few typical stars for each main subtype. GSC 05536-00897 is a DSCT pulsating star with a very short pulsation period of about 0.04562 day (1 hour and 5 minutes!) and a very small amplitude of the light curve variation between magnitude 13.92 and 13.98 CV. Data from CRTS survey were available for this star and added to our data. The resulting light curve is quite symmetric and there is no evidence of amplitude and/or period variation, at least compared to the old data from CRTS survey. Figure 5 shows the light curve phased with the main period of the pulsator.

CMC15 J163041.4-080658 is a DSCT pulsating star with a very short pulsation period of about 0.062443 day (1 hour and 29 minutes) and a large amplitude of the light curve variation between magnitude 16.22 and 16.78 CV. Data from CRTS survey were available for this star and were added to our data. The resulting light curve shows a rapid ascending branch and there is no evidence of amplitude and/or period variation, at least compared to the old data from CRTS survey. Figure 6 shows the light curve phased with the main period of the pulsator.

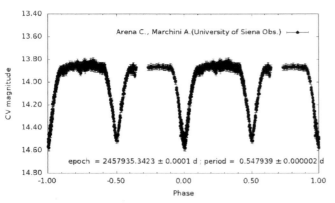

Figure 2. Folded light curve of GSC 05117-01301.

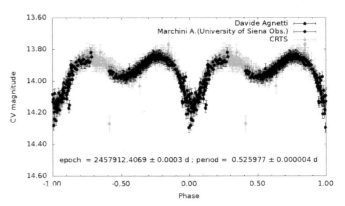

Figure 3. Folded light curve of GSC 05627-00248.

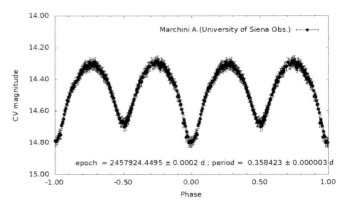

Figure 4. Folded light curve of UCAC4 370-097050.

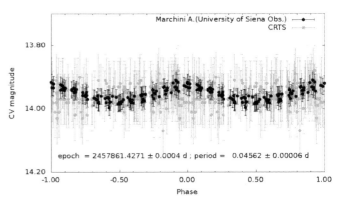

Figure 5. Folded light curve of GSC 05536-00897.

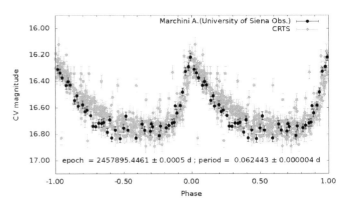

Figure 6. Folded light curve of CMC15 J163041.4-080658.

4. Conclusions

Mentoring the students in astronomy lab activities using a telescope with a CCD camera at the Astronomical Observatory of the University of Siena allowed us to collect a large amount of CCD images and dig inside this mine to search for new variables. Variables discovered this way are added to the AAVSO International Variable Star Index (VSX), to share them with the larger community of professional and amateur astronomers. In 2017 we discovered 24 new variable stars, specifically, 16 eclipsing binaries and 8 short period pulsators. The details of each of the new variable stars are given in Table 2 in order of increasing Right Ascension. Phase plots are shown in Figures 1 through 6 in section 3.

5. Acknowledgements

The authors firstly want to thank here Sebastián Otero, one of the VSX moderators, who kindly and eagerly helped us during the submission process with most valuable suggestions that were often crucial.

This work has made use of the VizieR catalog access tool, CDS, Strasbourg, France, the ASAS catalog, the CRTS catalog, the NSVS catalog, and of course the International Variable Star Index (VSX) operated by the AAVSO.

This publication makes use of data products from the Two Micron All Sky Survey (Skrutskie *et al.* 2006), which is a joint project of the University of Massachusetts and the Infrared Processing and Analysis Center/California Institute of Technology, funded by the National Aeronautics and Space Administration and the National Science Foundation.

This work has made use of data from the European Space Agency (ESA) mission Gaia (https://www.cosmos.esa.int/gaia), processed by the Gaia Data Processing and Analysis Consortium (DPAC, https://www.cosmos.esa.int/web/gaia/dpac/consortium). Funding for the DPAC has been provided by national institutions, in particular the institutions participating in the Gaia Multilateral Agreement.

Finally, we acknowledge with thanks the variable star observations from the AAVSO International Database contributed by observers worldwide and used in this research.

References

Drake, A. J. *et al.* 2014, *Astrophys. J., Suppl. Ser.*, **213**, 9.

DSFTA: University of Siena, Department of Physical Sciences, Earth and Environment. 2018 (http://www.dsfta.unisi.it).

Gaia Collaboration: Prusti, T., *et al.* 2016, *Astron. Astrophys.*, **595A**, 1 (Gaia Data Release 1).

Gaia Collaboration: Brown, A. G. A., *et al.* 2018, Gaia Data Release 2: Summary of the contents and survey properties, *Astron. Astrophys.*, special issue for Gaia DR2 (arXiv:1804.09365).

Lindegren, L., *et al.* 2018, Gaia Data Release 2: The astrometric solution, *Astron. Astrophys.*, special issue for Gaia DR2 (arXiv:1804.09366).

Liu, Q. Y., and Yang, Y. L. 2003, *Chinese J. Astron. Astrophys.*, **3**, 142.

O'Connell, D. J. K. 1951, *Riverview Coll. Obs. Publ.*, **2**, 85.

Papini, R., Franco, L., Marchini, A., and Salvaggio, F. 2015, *J. Amer. Assoc. Var. Star Obs.*, **43**, 207.

Papini, R., *et al.* 2017, *J. Amer. Assoc. Var. Star Obs.*, **45**, 219.

Pojmański, G. 2002, *Acta Astron.*, **52**, 397.

Skrutskie, M. F., *et al.* 2006, *Astron. J.*, **131**, 1163.

SSV-UAI: Unione Astrofili Italiani-Sezione Stelle Variabili. 2018 (http://stellevariabili.uai.it).

Watson, C., Henden, A. A., and Price, C. A. 2014, AAVSO International Variable Star Index VSX (Watson+, 2006–2017; http://www.aavso.org/vsx).

Wozniak, P., *et al.* 2004, *Astron. J.*, **127**, 2436.

Visual Times of Maxima for Short Period Pulsating Stars IV

Gerard Samolyk

P.O. Box 20677, Greenfield, WI 53220; gsamolyk@wi.rr.com

Received October 1, 2018; accepted October 2, 2018

Abstract   This compilation contains 556 times of maxima of 8 short period pulsating stars (primarily RR Lyrae type): TW Her, VX Her, AR Her, DY Her, SZ Hya, UU Hya, DG Hya, DH Hya. These were reduced from a portion of the visual observations made from 1966 to 2014 that are included in the AAVSO International Database.

1. Observations

This is the fourth in a series of papers to publish of times of maxima derived from visual observations reported to the AAVSO International Database as part of the AAVSO RR Lyr Committee legacy program. The goal of this project is to fill some historical gaps in the O–C history for these stars. This list contains times of maxima for RR Lyr stars located in the constellations Hercules and Hydra. This list will be web-archived and made available through the AAVSO ftp site at ftp://ftp.aavso.org/public/datasets/gsamj462vismax4.txt

These observations were reduced by the writer using the PERANSO program (Vanmunster 2007). The linear elements in the *General Catalogue of Variable Stars* (Kholopov *et al.* 1985) were used to compute the O–C values for all stars.

Figures 1, 2, and 3 are O–C plots for three of the stars included in Table 1. These plots include the visual times of maxima listed in this paper plus more recent times of maxima observed with CCDs. The circled CCD times of maxima on the plots were previously published in *JAAVSO* (Samolyk 2010–2018).

References

Kholopov, P. N., *et al.* 1985, *General Catalogue of Variable Stars*, 4th ed., Moscow.

Samolyk, G. 2010, *J. Amer. Assoc. Var. Star Obs.*, **38**, 12.

Samolyk, G. 2011, *J. Amer. Assoc. Var. Star Obs.*, **39**, 23.

Samolyk, G. 2012, *J. Amer. Assoc. Var. Star Obs.*, **40**, 923.

Samolyk, G. 2013, *J. Amer. Assoc. Var. Star Obs.*, **41**, 85.

Samolyk, G. 2014, *J. Amer. Assoc. Var. Star Obs.*, **42**, 124.

Samolyk, G. 2015, *J. Amer. Assoc. Var. Star Obs.*, **43**, 74.

Samolyk, G. 2016, *J. Amer. Assoc. Var. Star Obs.*, **44**, 66.

Samolyk, G. 2017, *J. Amer. Assoc. Var. Star Obs.*, **45**, 116.

Samolyk, G. 2018, *J. Amer. Assoc. Var. Star Obs.*, **46**, 74.

Vanmunster, T. 2007, PERANSO period analysis software (http://www.peranso.com).

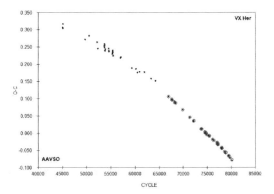

Figure 1. O–C plot for VX Her. The fundamental period of this star has been slowly decreasing since 1974.

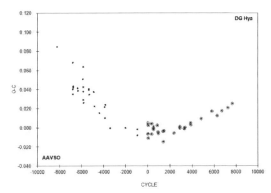

Figure 3. O–C plot for DG Hya. The fundamental period of this star has been increasing since 1985.

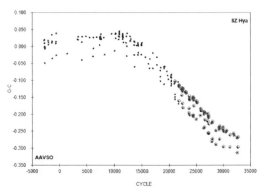

Figure 2. O–C plot for SZ Hya. There have been two significant changes in the fundamental period of this star since 1966.

Table 1. Recent times of minima of stars in the AAVSO short period pulsator program.

Star	JD (max) Hel. 2400000+	Cycle	O–C (day)	Observer	Error (day)	Star	JD (max) Hel. 2400000+	Cycle	O–C (day)	Observer	Error (day)
TW Her	44727.634	58014	−0.001	M. Baldwin	0.003	TW Her	53992.349	81199	−0.014	S. Swierczynski	0.002
TW Her	44866.698	58362	0.002	M. Heifner	0.003	TW Her	53996.348	81209	−0.011	S. Swierczynski	0.004
TW Her	44870.689	58372	−0.002	M. Heifner	0.003	TW Her	54002.340	81224	−0.013	S. Swierczynski	0.005
TW Her	45465.699	59861	0.003	M. Baldwin	0.003	TW Her	54008.334	81239	−0.013	S. Swierczynski	0.003
TW Her	45493.673	59931	0.006	M. Baldwin	0.004	TW Her	54682.460	82926	−0.013	S. Swierczynski	0.003
TW Her	45509.652	59971	0.000	G. Chaple	0.004	VX Her	42230.727	44974	0.307	H. Smith	0.004
TW Her	45511.650	59976	0.000	G. Chaple	0.003	VX Her	42236.645	44987	0.305	H. Smith	0.004
TW Her	45515.650	59986	0.004	G. Chaple	0.004	VX Her	42241.667	44998	0.318	H. Smith	0.003
TW Her	46173.796	61633	0.009	M. Baldwin	0.004	VX Her	44368.668	49669	0.273	M. Baldwin	0.003
TW Her	46181.782	61653	0.002	M. Baldwin	0.003	VX Her	44746.638	50499	0.283	M. Baldwin	0.003
TW Her	46193.768	61683	0.001	M. Baldwin	0.005	VX Her	45490.699	52133	0.265	M. Baldwin	0.006
TW Her	46205.751	61713	−0.004	M. Baldwin	0.003	VX Her	45562.630	52291	0.247	G. Chaple	0.004
TW Her	46211.749	61728	0.000	M. Baldwin	0.004	VX Her	46142.782	53565	0.254	M. Baldwin	0.004
TW Her	46233.732	61783	0.005	M. Baldwin	0.003	VX Her	46173.748	53633	0.255	M. Baldwin	0.006
TW Her	46275.688	61888	0.003	M. Baldwin	0.005	VX Her	46178.752	53644	0.250	M. Baldwin	0.006
TW Her	46289.676	61923	0.005	M. Baldwin	0.007	VX Her	46194.700	53679	0.260	M. Baldwin	0.005
TW Her	46325.644	62013	0.008	M. Heifner	0.003	VX Her	46210.619	53714	0.240	M. Baldwin	0.004
TW Her	46329.643	62023	0.011	M. Baldwin	0.004	VX Her	46239.767	53778	0.245	M. Baldwin	0.005
TW Her	46916.647	63492	0.003	M. Baldwin	0.002	VX Her	46571.736	54507	0.247	R. Hill	0.004
TW Her	46939.827	63550	0.007	P. Atwood	0.005	VX Her	46591.766	54551	0.240	R. Hill	0.006
TW Her	47676.680	65394	−0.004	M. Baldwin	0.002	VX Her	46602.692	54575	0.238	R. Hill	0.004
TW Her	47788.573	65674	0.002	M. Baldwin	0.003	VX Her	46914.616	55260	0.232	M. Baldwin	0.003
TW Her	47790.572	65679	0.003	M. Baldwin	0.002	VX Her	46939.668	55315	0.238	M. Baldwin	0.004
TW Her	47794.569	65689	0.004	M. Baldwin	0.003	VX Her	46944.674	55326	0.235	M. Baldwin	0.003
TW Her	47796.566	65694	0.003	M. Baldwin	0.004	VX Her	46947.868	55333	0.241	P. Atwood	0.004
TW Her	47804.550	65714	−0.005	M. Baldwin	0.003	VX Her	46968.800	55379	0.226	R. Hill	0.005
TW Her	47808.554	65724	0.002	M. Baldwin	0.003	VX Her	47678.720	56938	0.220	R. Hill	0.004
TW Her	47810.557	65729	0.008	M. Baldwin	0.003	VX Her	47703.768	56993	0.222	R. Hill	0.005
TW Her	48006.752	66220	−0.001	M. Baldwin	0.003	VX Her	48744.718	59279	0.190	M. Baldwin	0.003
TW Her	48188.568	66675	−0.003	M. Baldwin	0.003	VX Her	49117.666	60098	0.188	M. Baldwin	0.005
TW Her	48190.568	66680	−0.001	M. Baldwin	0.002	VX Her	49250.624	60390	0.177	M. Baldwin	0.004
TW Her	48194.566	66690	0.001	M. Baldwin	0.005	VX Her	49474.670	60882	0.179	M. Baldwin	0.005
TW Her	48196.571	66695	0.008	M. Baldwin	0.004	VX Her	49928.675	61879	0.177	M. Baldwin	0.004
TW Her	48202.554	66710	−0.003	M. Baldwin	0.003	VX Her	50539.767	63221	0.160	M. Baldwin	0.003
TW Her	48208.554	66725	0.003	M. Baldwin	0.003	VX Her	50957.793	64139	0.153	R. Hill	0.005
TW Her	48210.545	66730	−0.004	M. Baldwin	0.006	VX Her	52494.621	67514	0.098	R. Berg	0.006
TW Her	48414.743	67241	−0.001	M. Baldwin	0.002	VX Her	52556.552	67650	0.098	R. Berg	0.003
TW Her	48444.715	67316	0.000	M. Baldwin	0.003	VX Her	52757.823	68092	0.094	R. Hill	0.004
TW Her	48452.709	67336	0.002	M. Baldwin	0.003	VX Her	52812.462	68212	0.089	T. Fabjan	0.005
TW Her	48454.702	67341	−0.003	M. Baldwin	0.004	AR Her	39668.702	−3798	−0.479	M. Baldwin	0.006
TW Her	48480.673	67406	−0.006	M. Baldwin	0.003	AR Her	44431.950	6336	−0.494	G. Samolyk	0.007
TW Her	48482.673	67411	−0.003	M. Baldwin	0.004	AR Her	44792.515	7103	−0.441	G. Samolyk	0.005
TW Her	48484.673	67416	−0.002	M. Baldwin	0.003	AR Her	46174.735	10044	−0.573	M. Baldwin	0.006
TW Her	48508.652	67476	0.001	M. Baldwin	0.003	AR Her	46181.744	10059	−0.615	M. Baldwin	0.006
TW Her	48526.628	67521	−0.005	M. Baldwin	0.005	AR Her	46206.668	10112	−0.602	M. Baldwin	0.007
TW Her	48546.616	67571	0.004	M. Baldwin	0.004	AR Her	46211.832	10123	−0.608	M. Baldwin	0.005
TW Her	48894.662	68442	−0.003	M. Baldwin	0.003	AR Her	46247.626	10199	−0.537	G. Samolyk	0.003
TW Her	49129.624	69030	−0.005	M. Baldwin	0.003	AR Her	46253.709	10212	−0.564	M. Baldwin	0.006
TW Her	49133.627	69040	0.002	M. Baldwin	0.005	AR Her	46270.638	10248	−0.556	G. Samolyk	0.006
TW Her	49857.698	70852	−0.003	M. Baldwin	0.004	AR Her	46324.675	10363	−0.572	M. Baldwin	0.004
TW Her	49859.699	70857	0.000	M. Baldwin	0.004	AR Her	46325.601	10365	−0.586	M. Baldwin	0.006
TW Her	49873.681	70892	−0.004	M. Baldwin	0.004	AR Her	46333.590	10382	−0.588	M. Baldwin	0.005
TW Her	49901.653	70962	−0.003	M. Baldwin	0.003	AR Her	46348.593	10414	−0.626	M. Baldwin	0.007
TW Her	49953.605	71092	0.000	M. Baldwin	0.003	AR Her	46511.721	10761	−0.597	M. Baldwin	0.006
TW Her	49955.602	71097	0.000	M. Baldwin	0.004	AR Her	46520.666	10780	−0.583	M. Baldwin	0.003
TW Her	49965.589	71122	−0.003	M. Baldwin	0.005	AR Her	46526.755	10793	−0.604	M. Baldwin	0.006
TW Her	50351.607	72088	0.001	M. Baldwin	0.003	AR Her	46527.682	10795	−0.617	M. Baldwin	0.003
TW Her	50539.813	72559	−0.005	M. Baldwin	0.005	AR Her	46606.693	10963	−0.571	G. Samolyk	0.006
TW Her	50541.814	72564	−0.002	M. Baldwin	0.004	AR Her	46654.598	11065	−0.609	G. Samolyk	0.007
TW Her	50957.793	73605	−0.007	R. Hill	0.004	AR Her	46659.742	11076	−0.635	G. Samolyk	0.003
TW Her	51021.733	73765	−0.003	R. Berg	0.004	AR Her	46678.584	11116	−0.594	G. Samolyk	0.003
TW Her	51025.721	73775	−0.011	R. Berg	0.003	AR Her	46701.631	11165	−0.579	M. Baldwin	0.007
TW Her	51318.630	74508	−0.008	M. Baldwin	0.003	AR Her	46709.597	11182	−0.603	G. Samolyk	0.004
TW Her	51397.754	74706	−0.005	M. Baldwin	0.003	AR Her	46725.541	11216	−0.640	M. Baldwin	0.006
TW Her	51439.704	74811	−0.013	M. Baldwin	0.005	AR Her	46831.807	11442	−0.600	M. Baldwin	0.003
TW Her	53990.353	81194	−0.011	S. Swierczynski	0.004	AR Her	46888.672	11563	−0.609	M. Baldwin	0.007

Table continued on following pages

Table 1. Recent times of minima of stars in the AAVSO short period pulsator program, cont.

Star	JD (max) Hel. 2400000+	Cycle	O–C (day)	Observer	Error (day)	Star	JD (max) Hel. 2400000+	Cycle	O–C (day)	Observer	Error (day)
AR Her	46905.589	11599	–0.613	M. Baldwin	0.006	AR Her	49965.652	18110	–0.902	M. Baldwin	0.003
AR Her	46910.729	11610	–0.643	M. Baldwin	0.004	AR Her	50006.529	18197	–0.918	G. Samolyk	0.003
AR Her	46911.702	11612	–0.610	M. Baldwin	0.006	AR Her	50284.749	18789	–0.954	G. Samolyk	0.002
AR Her	46912.608	11614	–0.644	M. Baldwin	0.003	AR Her	50286.636	18793	–0.947	G. Chaple	0.002
AR Her	46920.621	11631	–0.622	G. Samolyk	0.005	AR Her	50301.712	18825	–0.912	M. Baldwin	0.005
AR Her	46935.661	11663	–0.623	M. Baldwin	0.005	AR Her	50302.642	18827	–0.922	M. Baldwin	0.004
AR Her	46942.683	11678	–0.651	M. Baldwin	0.006	AR Her	50326.591	18878	–0.945	M. Baldwin	0.003
AR Her	46948.829	11691	–0.615	M. Baldwin	0.005	AR Her	50542.820	19338	–0.928	M. Baldwin	0.008
AR Her	46951.641	11697	–0.624	G. Samolyk	0.002	AR Her	50575.676	19408	–0.974	M. Baldwin	0.005
AR Her	46974.674	11746	–0.622	M. Baldwin	0.005	AR Her	50614.686	19491	–0.977	G. Samolyk	0.004
AR Her	46997.731	11795	–0.596	M. Baldwin	0.005	AR Her	50726.567	19729	–0.962	G. Samolyk	0.003
AR Her	47022.641	11848	–0.598	M. Baldwin	0.004	AR Her	50950.768	20206	–0.965	G. Samolyk	0.003
AR Her	47037.639	11880	–0.641	M. Baldwin	0.003	AR Her	50957.803	20221	–0.980	R. Hill	0.006
AR Her	47038.573	11882	–0.647	M. Baldwin	0.005	AR Her	50965.768	20238	–1.006	R. Hill	0.008
AR Her	47086.552	11984	–0.611	M. Baldwin	0.004	AR Her	50967.668	20242	–0.986	R. Berg	0.006
AR Her	47232.685	12295	–0.656	M. Baldwin	0.006	AR Her	50981.780	20272	–0.975	R. Hill	0.007
AR Her	47241.651	12314	–0.621	M. Baldwin	0.005	AR Her	50991.634	20293	–0.991	M. Baldwin	0.005
AR Her	47248.699	12329	–0.623	M. Baldwin	0.003	AR Her	51005.762	20323	–0.964	R. Berg	0.004
AR Her	47264.638	12363	–0.665	M. Baldwin	0.003	AR Her	51006.693	20325	–0.973	G. Samolyk	0.002
AR Her	47271.712	12378	–0.642	M. Baldwin	0.005	AR Her	51006.714	20325	–0.952	R. Berg	0.006
AR Her	47295.671	12429	–0.654	G. Samolyk	0.004	AR Her	51007.642	20327	–0.964	M. Baldwin	0.003
AR Her	47308.825	12457	–0.661	R. Hill	0.006	AR Her	51012.795	20338	–0.981	G. Samolyk	0.004
AR Her	47325.747	12493	–0.660	M. Baldwin	0.004	AR Her	51021.698	20357	–1.009	R. Berg	0.005
AR Her	47358.633	12563	–0.676	M. Baldwin	0.004	AR Her	51045.708	20408	–0.970	R. Berg	0.005
AR Her	47382.637	12614	–0.643	M. Baldwin	0.004	AR Her	51046.635	20410	–0.983	M. Baldwin	0.004
AR Her	47390.608	12631	–0.663	M. Baldwin	0.006	AR Her	51069.670	20459	–0.980	G. Samolyk	0.004
AR Her	47406.620	12665	–0.632	M. Baldwin	0.004	AR Her	51069.678	20459	–0.972	M. Baldwin	0.005
AR Her	47632.691	13146	–0.644	G. Samolyk	0.005	AR Her	51109.598	20544	–1.004	G. Samolyk	0.004
AR Her	47670.743	13227	–0.664	R. Hill	0.004	AR Her	51319.697	20991	–1.008	M. Baldwin	0.006
AR Her	47685.808	13259	–0.640	M. Baldwin	0.009	AR Her	51421.709	21208	–0.992	M. Baldwin	0.004
AR Her	47687.674	13263	–0.654	M. Baldwin	0.004	AR Her	51423.591	21212	–0.990	M. Baldwin	0.004
AR Her	47688.621	13265	–0.647	M. Baldwin	0.006	AR Her	51428.751	21223	–1.000	M. Baldwin	0.006
AR Her	47712.610	13316	–0.630	G. Samolyk	0.003	AR Her	51429.681	21225	–1.010	M. Baldwin	0.005
AR Her	47718.701	13329	–0.649	M. Baldwin	0.005	AR Her	51437.645	21242	–1.037	M. Baldwin	0.005
AR Her	47733.723	13361	–0.668	M. Baldwin	0.004	AR Her	51486.568	21346	–0.997	M. Baldwin	0.004
AR Her	47773.661	13446	–0.682	M. Baldwin	0.005	AR Her	51657.607	21710	–1.048	M. Baldwin	0.006
AR Her	47790.602	13482	–0.662	G. Samolyk	0.003	AR Her	51664.661	21725	–1.044	M. Baldwin	0.005
AR Her	47790.611	13482	–0.653	M. Baldwin	0.005	AR Her	51805.651	22025	–1.063	G. Samolyk	0.006
AR Her	47798.594	13499	–0.661	G. Samolyk	0.005	AR Her	51813.643	22042	–1.061	G. Samolyk	0.006
AR Her	47807.547	13518	–0.639	M. Baldwin	0.004	AR Her	51814.594	22044	–1.050	M. Baldwin	0.005
AR Her	47978.619	13882	–0.657	M. Baldwin	0.005	AR Her	52471.693	23442	–1.050	R. Berg	0.003
AR Her	48000.713	13929	–0.654	M. Baldwin	0.005	AR Her	52487.657	23476	–1.067	R. Berg	0.004
AR Her	48048.654	14031	–0.656	R. Hill	0.006	AR Her	52494.697	23491	–1.078	R. Berg	0.004
AR Her	48415.644	14812	–0.758	M. Baldwin	0.005	AR Her	55387.486	29646	–1.311	J. Starzomski	0.004
AR Her	48421.806	14825	–0.706	M. Baldwin	0.004	DY Her	39672.643	41937	0.003	M. Baldwin	0.003
AR Her	48445.740	14876	–0.744	M. Baldwin	0.005	DY Her	39672.791	41938	0.003	M. Baldwin	0.005
AR Her	48452.836	14891	–0.698	M. Baldwin	0.003	DY Her	39686.758	42032	–0.002	M. Baldwin	0.005
AR Her	48526.607	15048	–0.721	M. Baldwin	0.004	DY Her	44771.727	76244	–0.008	M. Heifner	0.002
AR Her	48744.674	15512	–0.747	M. Baldwin	0.005	DY Her	44778.713	76291	–0.008	M. Heifner	0.002
AR Her	48773.814	15574	–0.749	G. Samolyk	0.005	DY Her	45465.690	80913	–0.005	M. Baldwin	0.006
AR Her	48822.659	15678	–0.787	M. Baldwin	0.003	DY Her	45520.684	81283	–0.008	M. Heifner	0.004
AR Her	48885.630	15812	–0.800	G. Samolyk	0.004	DY Her	45535.697	81384	–0.004	M. Heifner	0.002
AR Her	48893.623	15829	–0.797	G. Samolyk	0.003	DY Her	45556.651	81525	–0.007	G. Chaple	0.002
AR Her	49213.687	16510	–0.822	M. Baldwin	0.005	DY Her	46194.720	85818	–0.012	M. Baldwin	0.006
AR Her	49254.601	16597	–0.801	G. Samolyk	0.003	DY Her	46211.826	85933	0.001	M. Baldwin	0.005
AR Her	49278.547	16648	–0.826	G. Samolyk	0.004	DY Her	46944.718	90864	–0.008	M. Baldwin	0.003
AR Her	49423.782	16957	–0.830	M. Baldwin	0.005	DY Her	46948.730	90891	–0.009	M. Baldwin	0.003
AR Her	49480.659	17078	–0.826	G. Samolyk	0.003	DY Her	46973.701	91059	–0.008	M. Baldwin	0.003
AR Her	49614.571	17363	–0.872	M. Baldwin	0.004	DY Her	46974.746	91066	–0.003	M. Baldwin	0.003
AR Her	49637.621	17412	–0.854	G. Samolyk	0.004	DY Her	47027.647	91422	–0.015	M. Baldwin	0.004
AR Her	49832.670	17827	–0.866	M. Baldwin	0.003	DY Her	47388.684	93851	–0.004	M. Baldwin	0.004
AR Her	49900.786	17972	–0.904	M. Baldwin	0.005	DY Her	47412.609	94012	–0.008	M. Baldwin	0.002
AR Her	49901.729	17974	–0.901	M. Baldwin	0.004	DY Her	47676.722	95789	–0.013	M. Baldwin	0.004
AR Her	49918.672	18010	–0.879	M. Baldwin	0.008	DY Her	47683.715	95836	–0.006	M. Baldwin	0.005
AR Her	49926.662	18027	–0.880	M. Baldwin	0.006	DY Her	47748.663	96273	–0.010	M. Baldwin	0.004
AR Her	49957.674	18093	–0.890	M. Baldwin	0.005	DY Her	48004.748	97996	–0.017	M. Baldwin	0.004

Table continued on following pages

Table 1. Recent times of minima of stars in the AAVSO short period pulsator program, cont.

Star	JD (max) Hel. 2400000+	Cycle	O–C (day)	Observer	Error (day)	Star	JD (max) Hel. 2400000+	Cycle	O–C (day)	Observer	Error (day)
DY Her	48061.680	98379	−0.010	M. Baldwin	0.003	SZ Hya	46517.640	10867	0.039	M. Baldwin	0.003
DY Her	48065.687	98406	−0.016	M. Baldwin	0.002	SZ Hya	46518.722	10869	0.046	M. Baldwin	0.005
DY Her	48067.775	98420	−0.009	M. Baldwin	0.003	SZ Hya	46532.688	10895	0.044	M. Baldwin	0.006
DY Her	48151.600	98984	−0.012	M. Baldwin	0.003	SZ Hya	46829.771	11448	0.033	M. Baldwin	0.003
DY Her	49076.835	105209	−0.008	M. Baldwin	0.006	SZ Hya	46835.671	11459	0.023	G. Samolyk	0.003
DY Her	49161.696	105780	−0.015	M. Baldwin	0.006	SZ Hya	46850.727	11487	0.037	M. Baldwin	0.004
DY Her	49482.742	107940	−0.013	M. Baldwin	0.003	SZ Hya	46857.705	11500	0.030	M. Baldwin	0.005
DY Her	49488.692	107980	−0.008	M. Baldwin	0.006	SZ Hya	46857.713	11500	0.038	G. Samolyk	0.002
DY Her	49859.675	110476	−0.009	M. Baldwin	0.004	SZ Hya	46858.782	11502	0.033	M. Baldwin	0.004
DY Her	49868.738	110537	−0.012	M. Baldwin	0.003	SZ Hya	46878.655	11539	0.028	M. Baldwin	0.004
DY Her	51436.647	121086	−0.016	M. Baldwin	0.004	SZ Hya	46914.635	11606	0.013	M. Baldwin	0.003
DY Her	52487.622	128157	−0.013	R. Berg	0.003	SZ Hya	47161.765	12066	0.013	G. Samolyk	0.004
DY Her	52489.700	128171	−0.016	R. Berg	0.006	SZ Hya	47219.805	12174	0.031	G. Samolyk	0.003
SZ Hya	39140.775	−2864	0.019	M. Baldwin	0.002	SZ Hya	47231.596	12196	0.002	M. Baldwin	0.004
SZ Hya	39148.815	−2849	0.000	M. Baldwin	0.006	SZ Hya	47232.642	12198	−0.026	M. Baldwin	0.005
SZ Hya	39168.704	−2812	0.011	M. Baldwin	0.006	SZ Hya	47267.629	12263	0.040	M. Baldwin	0.004
SZ Hya	39169.788	−2810	0.021	M. Baldwin	0.005	SZ Hya	47557.730	12803	0.031	M. Baldwin	0.004
SZ Hya	39182.613	−2786	−0.048	M. Baldwin	0.006	SZ Hya	47558.791	12805	0.018	M. Baldwin	0.004
SZ Hya	39197.705	−2758	0.002	M. Baldwin	0.007	SZ Hya	47585.653	12855	0.018	M. Baldwin	0.005
SZ Hya	39204.682	−2745	−0.006	M. Baldwin	0.006	SZ Hya	47621.653	12922	0.023	M. Baldwin	0.006
SZ Hya	39225.637	−2706	−0.003	M. Baldwin	0.006	SZ Hya	47948.786	13531	−0.023	M. Baldwin	0.006
SZ Hya	39530.811	−2138	0.019	M. Baldwin	0.003	SZ Hya	47952.586	13538	0.016	G. Samolyk	0.002
SZ Hya	39556.593	−2090	0.013	M. Baldwin	0.002	SZ Hya	47954.724	13542	0.005	M. Baldwin	0.006
SZ Hya	39558.735	−2086	0.006	M. Baldwin	0.004	SZ Hya	47976.763	13583	0.017	M. Baldwin	0.008
SZ Hya	39890.755	−1468	0.012	M. Baldwin	0.004	SZ Hya	47997.639	13622	−0.059	M. Baldwin	0.005
SZ Hya	39896.688	−1457	0.035	M. Baldwin	0.007	SZ Hya	48004.697	13635	0.015	M. Baldwin	0.006
SZ Hya	39912.736	−1427	−0.034	M. Baldwin	0.007	SZ Hya	48216.897	14030	0.005	M. Baldwin	0.006
SZ Hya	39918.696	−1416	0.016	M. Baldwin	0.004	SZ Hya	48320.592	14223	0.012	M. Baldwin	0.005
SZ Hya	40293.689	−718	0.015	M. Baldwin	0.008	SZ Hya	48335.599	14251	−0.023	G. Samolyk	0.004
SZ Hya	40294.758	−716	0.010	M. Baldwin	0.005	SZ Hya	48357.633	14292	−0.016	M. Baldwin	0.006
SZ Hya	40321.650	−666	0.040	M. Baldwin	0.007	SZ Hya	48357.652	14292	0.003	G. Samolyk	0.004
SZ Hya	40323.734	−662	−0.025	M. Baldwin	0.007	SZ Hya	48379.684	14333	0.008	M. Baldwin	0.006
SZ Hya	41393.384	1329	−0.020	M. Baldwin	0.006	SZ Hya	48654.721	14845	−0.022	G. Samolyk	0.003
SZ Hya	42429.763	3258	0.022	M. Baldwin	0.004	SZ Hya	48661.734	14858	0.007	G. Samolyk	0.003
SZ Hya	42477.585	3347	0.030	M. Baldwin	0.004	SZ Hya	48682.684	14897	0.004	M. Baldwin	0.005
SZ Hya	42507.661	3403	0.021	M. Baldwin	0.003	SZ Hya	48683.764	14899	0.010	M. Baldwin	0.008
SZ Hya	42845.579	4032	0.014	M. Baldwin	0.002	SZ Hya	48718.687	14964	0.012	M. Baldwin	0.004
SZ Hya	42861.643	4062	−0.039	M. Baldwin	0.002	SZ Hya	49064.627	15608	−0.030	G. Samolyk	0.005
SZ Hya	42874.605	4086	0.029	M. Baldwin	0.003	SZ Hya	49333.765	16109	−0.050	G. Samolyk	0.002
SZ Hya	43242.616	4771	0.031	M. Baldwin	0.003	SZ Hya	49433.678	16295	−0.063	M. Baldwin	0.005
SZ Hya	43610.622	5456	0.027	M. Baldwin	0.004	SZ Hya	49483.669	16388	−0.036	M. Baldwin	0.006
SZ Hya	43935.597	6061	−0.028	M. Baldwin	0.003	SZ Hya	49778.586	16937	−0.064	M. Baldwin	0.006
SZ Hya	43970.573	6126	0.027	G. Samolyk	0.003	SZ Hya	49787.765	16954	−0.018	M. Baldwin	0.005
SZ Hya	43986.658	6156	−0.005	G. Samolyk	0.005	SZ Hya	50169.707	17665	−0.053	M. Baldwin	0.005
SZ Hya	44317.632	6772	0.029	M. Baldwin	0.004	SZ Hya	50190.683	17704	−0.030	M. Baldwin	0.004
SZ Hya	44608.793	7314	0.006	G. Samolyk	0.004	SZ Hya	50488.793	18259	−0.088	G. Samolyk	0.003
SZ Hya	44622.779	7340	0.024	G. Samolyk	0.003	SZ Hya	50492.572	18266	−0.070	G. Samolyk	0.005
SZ Hya	44629.766	7353	0.027	G. Samolyk	0.002	SZ Hya	50514.567	18307	−0.102	G. Samolyk	0.003
SZ Hya	44629.768	7353	0.029	G. Hanson	0.004	SZ Hya	50523.741	18324	−0.061	G. Samolyk	0.004
SZ Hya	44672.750	7433	0.031	G. Samolyk	0.004	SZ Hya	50869.726	18968	−0.058	M. Baldwin	0.006
SZ Hya	44686.724	7459	0.037	G. Hanson	0.003	SZ Hya	50870.796	8970	−0.063	R. Hill	0.006
SZ Hya	44700.680	7485	0.025	M. Heifner	0.005	SZ Hya	50876.659	18981	−0.110	R. Hill	0.006
SZ Hya	44736.672	7552	0.022	G. Samolyk	0.004	SZ Hya	50926.674	19074	−0.058	G. Samolyk	0.003
SZ Hya	44995.628	8034	0.028	G. Samolyk	0.003	SZ Hya	51160.896	19510	−0.073	G. Samolyk	0.003
SZ Hya	45060.584	8155	−0.022	G. Samolyk	0.003	SZ Hya	51223.741	19627	−0.085	G. Samolyk	0.007
SZ Hya	45753.685	9445	0.039	G. Samolyk	0.004	SZ Hya	51308.614	19785	−0.096	M. Baldwin	0.005
SZ Hya	46058.787	10013	−0.011	M. Baldwin	0.006	SZ Hya	51549.842	20234	−0.089	M. Baldwin	0.005
SZ Hya	46114.709	10117	0.038	M. Baldwin	0.006	SZ Hya	51583.674	20297	−0.103	M. Baldwin	0.004
SZ Hya	46142.647	10169	0.039	M. Baldwin	0.005	SZ Hya	51603.571	20334	−0.084	M. Baldwin	0.005
SZ Hya	46142.648	10169	0.040	G. Samolyk	0.004	SZ Hya	51606.764	20340	−0.114	R. Hill	0.006
SZ Hya	46143.708	10171	0.026	M. Baldwin	0.005	SZ Hya	51611.632	20349	−0.081	M. Baldwin	0.005
SZ Hya	46150.705	10184	0.039	M. Baldwin	0.006	SZ Hya	51633.641	20390	−0.099	M. Baldwin	0.004
SZ Hya	46435.435	10714	0.031	T. Cooper	0.005	SZ Hya	51633.649	20390	−0.091	G. Samolyk	0.003
SZ Hya	46436.500	10716	0.022	T. Cooper	0.004	SZ Hya	51640.615	20403	−0.109	M. Baldwin	0.005
SZ Hya	46511.722	10856	0.030	G. Samolyk	0.002	SZ Hya	51930.725	20943	−0.109	G. Samolyk	0.004
SZ Hya	46511.728	10856	0.036	M. Baldwin	0.005	SZ Hya	51937.712	20956	−0.106	G. Samolyk	0.005

Table continued on following pages

Table 1. Recent times of minima of stars in the AAVSO short period pulsator program, cont.

Star	JD (max) Hel. 2400000+	Cycle	O–C (day)	Observer	Error (day)	Star	JD (max) Hel. 2400000+	Cycle	O–C (day)	Observer	Error (day)
SZ Hya	51981.724	21038	–0.148	R. Hill	0.006	DG Hya	47586.671	–6348	0.039	M. Baldwin	0.004
SZ Hya	51988.701	21051	–0.155	R. Hill	0.007	DG Hya	47914.779	–5913	0.051	M. Baldwin	0.006
SZ Hya	51995.700	21064	–0.140	R. Hill	0.006	DG Hya	47942.673	–5876	0.038	M. Baldwin	0.009
SZ Hya	52319.693	21667	–0.103	M. Baldwin	0.004	DG Hya	47945.716	–5872	0.064	M. Baldwin	0.006
SZ Hya	52347.621	21719	–0.111	M. Baldwin	0.005	DG Hya	47948.698	–5868	0.029	M. Baldwin	0.006
SZ Hya	52356.698	21736	–0.167	R. Hill	0.006	DG Hya	47954.745	–5860	0.043	M. Baldwin	0.007
UU Hya	39178.795	–573	0.029	M. Baldwin	0.008	DG Hya	47976.602	–5831	0.026	M. Baldwin	0.007
UU Hya	39197.624	–537	–0.001	M. Baldwin	0.011	DG Hya	48320.550	–5375	0.039	M. Baldwin	0.007
UU Hya	39530.814	99	0.008	M. Baldwin	0.006	DG Hya	48335.636	–5355	0.041	M. Baldwin	0.005
UU Hya	39595.756	223	–0.010	M. Baldwin	0.007	DG Hya	48356.749	–5327	0.035	R. Hill	0.004
UU Hya	39912.666	828	–0.040	M. Baldwin	0.006	DG Hya	48654.678	–4932	0.038	M. Baldwin	0.008
UU Hya	39915.826	834	–0.023	M. Baldwin	0.005	DG Hya	48718.773	–4847	0.023	M. Baldwin	0.006
UU Hya	42832.726	6402	–0.023	M. Baldwin	0.004	DG Hya	49047.616	–4411	0.016	M. Baldwin	0.008
UU Hya	42843.734	6423	–0.016	M. Baldwin	0.004	DG Hya	49397.591	–3947	0.022	G. Samolyk	0.004
UU Hya	42844.788	6425	–0.010	M. Baldwin	0.004	DG Hya	49430.766	–3903	0.010	M. Baldwin	0.008
UU Hya	42863.631	6461	–0.026	M. Baldwin	0.008	DG Hya	49443.602	–3886	0.024	M. Baldwin	0.004
UU Hya	42874.645	6482	–0.013	M. Baldwin	0.005	DG Hya	49780.724	–3439	0.000	M. Baldwin	0.003
UU Hya	42886.702	6505	–0.005	M. Baldwin	0.005	DG Hya	50842.698	–2031	0.000	M. Baldwin	0.007
UU Hya	43610.679	7887	–0.015	M. Baldwin	0.004	DG Hya	51640.679	–973	–0.008	M. Baldwin	0.008
UU Hya	43631.609	7927	–0.039	M. Baldwin	0.004	DG Hya	51643.702	–969	–0.001	M. Baldwin	0.007
UU Hya	43960.622	8555	–0.015	M. Baldwin	0.007	DG Hya	52380.593	8	–0.006	M. Baldwin	0.006
UU Hya	44696.679	9960	0.007	M. Baldwin	0.005	DG Hya	52757.722	508	0.002	R. Hill	0.007
UU Hya	46114.775	12667	–0.009	M. Baldwin	0.005	DH Hya	39178.705	16365	0.005	M. Baldwin	0.004
UU Hya	46517.641	13436	0.002	M. Baldwin	0.003	DH Hya	39180.663	16369	0.007	M. Baldwin	0.003
UU Hya	46529.714	13459	0.026	M. Baldwin	0.008	DH Hya	39181.643	16371	0.009	M. Baldwin	0.005
UU Hya	46850.831	14072	0.012	M. Baldwin	0.005	DH Hya	39182.622	16373	0.010	M. Baldwin	0.005
UU Hya	46858.668	14087	–0.009	M. Baldwin	0.008	DH Hya	39197.772	16404	0.002	M. Baldwin	0.004
UU Hya	46881.753	14131	0.026	R. Hill	0.007	DH Hya	39200.709	16410	0.005	M. Baldwin	0.004
UU Hya	46912.632	14190	–0.003	M. Baldwin	0.004	DH Hya	39203.658	16416	0.020	M. Baldwin	0.006
UU Hya	47231.675	14799	0.003	M. Baldwin	0.005	DH Hya	39204.648	16418	0.032	M. Baldwin	0.008
UU Hya	47232.700	14801	–0.019	M. Baldwin	0.003	DH Hya	39225.646	16461	0.003	M. Baldwin	0.008
UU Hya	47241.637	14818	0.012	M. Baldwin	0.004	DH Hya	39528.831	17081	0.009	M. Baldwin	0.004
UU Hya	47243.741	14822	0.020	M. Baldwin	0.004	DH Hya	39530.783	17085	0.005	M. Baldwin	0.005
UU Hya	47594.707	15492	–0.005	M. Baldwin	0.005	DH Hya	39532.745	17089	0.011	M. Baldwin	0.004
UU Hya	47615.692	15532	0.025	R. Hill	0.009	DH Hya	39533.721	17091	0.009	M. Baldwin	0.003
UU Hya	47914.797	16103	0.001	M. Baldwin	0.004	DH Hya	39534.694	17093	0.004	M. Baldwin	0.002
UU Hya	47915.851	16105	0.007	M. Baldwin	0.005	DH Hya	39556.704	17138	0.009	M. Baldwin	0.004
UU Hya	47922.674	16118	0.020	M. Baldwin	0.002	DH Hya	39558.657	17142	0.006	M. Baldwin	0.003
UU Hya	47943.593	16158	–0.016	M. Baldwin	0.005	DH Hya	39582.618	17191	0.006	M. Baldwin	0.002
UU Hya	47954.621	16179	0.011	M. Baldwin	0.005	DH Hya	39886.780	17813	0.011	M. Baldwin	0.003
UU Hya	47955.669	16181	0.011	M. Baldwin	0.004	DH Hya	42491.667	23140	0.005	M. Baldwin	0.004
UU Hya	47976.615	16221	0.002	M. Baldwin	0.005	DH Hya	42843.744	23860	0.003	M. Baldwin	0.005
UU Hya	47977.643	16223	–0.017	M. Baldwin	0.006	DH Hya	42844.723	23862	0.004	M. Baldwin	0.004
UU Hya	47978.682	16225	–0.025	M. Baldwin	0.005	DH Hya	42845.706	23864	0.009	M. Baldwin	0.004
UU Hya	47999.674	16265	0.011	M. Baldwin	0.009	DH Hya	42871.616	23917	0.002	M. Baldwin	0.002
UU Hya	48000.725	16267	0.015	M. Baldwin	0.009	DH Hya	43226.622	24643	–0.005	M. Baldwin	0.003
UU Hya	48362.719	16958	0.015	M. Baldwin	0.006	DH Hya	43227.603	24645	–0.002	M. Baldwin	0.005
UU Hya	48658.686	17523	–0.003	M. Baldwin	0.005	DH Hya	43228.591	24647	0.008	M. Baldwin	0.005
UU Hya	48682.812	17569	0.025	M. Baldwin	0.004	DH Hya	43247.655	24686	0.001	M. Baldwin	0.003
UU Hya	48690.660	17584	0.016	M. Baldwin	0.005	DH Hya	43606.586	25420	0.008	M. Baldwin	0.005
UU Hya	48746.694	17691	–0.005	M. Baldwin	0.006	DH Hya	43960.624	26144	0.011	M. Baldwin	0.004
UU Hya	49417.759	18972	–0.015	M. Baldwin	0.004	DH Hya	43980.666	26185	0.004	M. Baldwin	0.004
UU Hya	49450.797	19035	0.019	M. Baldwin	0.006	DH Hya	43981.644	26187	0.004	M. Baldwin	0.005
UU Hya	49810.690	19722	0.014	M. Baldwin	0.006	DH Hya	43982.627	26189	0.009	M. Baldwin	0.004
UU Hya	50185.768	20438	0.002	R. Hill	0.006	DH Hya	44313.685	26866	0.015	M. Baldwin	0.007
UU Hya	50514.769	21066	0.014	M. Baldwin	0.004	DH Hya	44314.658	26868	0.010	M. Baldwin	0.005
UU Hya	50545.671	21125	0.008	M. Baldwin	0.007	DH Hya	44317.597	26874	0.015	M. Baldwin	0.004
UU Hya	50876.748	21757	0.000	R. Hill	0.006	DH Hya	46114.664	30549	0.014	M. Baldwin	0.004
UU Hya	51248.704	22467	0.010	M. Baldwin	0.005	DH Hya	46117.603	30555	0.019	M. Baldwin	0.006
DG Hya	46150.639	–8252	0.085	M. Baldwin	0.007	DH Hya	46490.705	31318	0.015	M. Baldwin	0.004
DG Hya	47227.648	–6824	0.035	M. Baldwin	0.008	DH Hya	46517.603	31373	0.018	M. Baldwin	0.003
DG Hya	47233.715	–6816	0.068	M. Baldwin	0.007	DH Hya	46845.721	32044	0.019	M. Baldwin	0.003
DG Hya	47245.755	–6800	0.040	R. Hill	0.006	DH Hya	46850.617	32054	0.025	M. Baldwin	0.004
DG Hya	47264.613	–6775	0.042	M. Baldwin	0.005	DH Hya	46914.664	32185	0.013	M. Baldwin	0.006
DG Hya	47267.631	–6771	0.043	M. Baldwin	0.007	DH Hya	46915.650	32187	0.021	G. Samolyk	0.004
DG Hya	47558.767	–6385	0.042	M. Baldwin	0.005	DH Hya	47204.643	32778	0.016	M. Baldwin	0.004

Table continued on next page

Table 1. Recent times of minima of stars in the AAVSO short period pulsator program, cont.

Star	JD (max) Hel. 2400000 +	Cycle	O–C (day)	Observer	Error (day)	Star	JD (max) Hel. 2400000 +	Cycle	O–C (day)	Observer	Error (day)
DH Hya	47226.661	32823	0.029	M. Baldwin	0.005	DH Hya	49801.726	38089	0.030	M. Baldwin	0.003
DH Hya	47271.633	32915	0.013	M. Baldwin	0.003	DH Hya	50110.781	38721	0.038	M. Baldwin	0.004
DH Hya	47531.790	33447	0.023	M. Baldwin	0.003	DH Hya	50138.652	38778	0.036	M. Baldwin	0.002
DH Hya	47604.654	33596	0.026	G. Samolyk	0.004	DH Hya	50158.706	38819	0.041	M. Baldwin	0.003
DH Hya	47955.759	34314	0.031	M. Baldwin	0.006	DH Hya	50182.652	38868	0.026	M. Baldwin	0.004
DH Hya	48004.649	34414	0.021	M. Baldwin	0.003	DH Hya	50514.698	39547	0.042	M. Baldwin	0.004
DH Hya	48648.660	35731	0.021	M. Baldwin	0.003	DH Hya	50842.808	40218	0.034	M. Baldwin	0.003
DH Hya	49018.834	36488	0.024	M. Baldwin	0.003	DH Hya	50843.782	40220	0.030	R. Hill	0.008
DH Hya	49397.812	37263	0.028	M. Baldwin	0.003	DH Hya	50869.711	40273	0.042	M. Baldwin	0.003
DH Hya	49401.728	37271	0.032	M. Baldwin	0.004	DH Hya	50872.641	40279	0.039	M. Baldwin	0.003
DH Hya	49423.730	37316	0.029	M. Baldwin	0.005	DH Hya	51248.687	41048	0.045	M. Baldwin	0.002
DH Hya	49428.624	37326	0.033	M. Baldwin	0.003	DH Hya	52757.731	44134	0.040	R. Hill	0.008
DH Hya	49778.739	38042	0.025	M. Baldwin	0.003	DH Hya	52758.723	44136	0.054	R. Hill	0.008

Recent Minima of 266 Eclipsing Binary Stars

Gerard Samolyk

P.O. Box 20677, Greenfield, WI 53220; gsamolyk@wi.rr.com

Received September 26, 2018; accepted September 26, 2018

Abstract   This paper continues the publication of times of minima for eclipsing binary stars from CCD observations reported to the AAVSO Eclipsing Binaries Section. Times of minima from observations received from February 2018 through August 2018 are presented.

1. Recent observations

The accompanying list contains times of minima calculated from recent CCD observations made by participants in the AAVSO's eclipsing binary program. This list will be web-archived and made available through the AAVSO ftp site at ftp://ftp.aavso.org/public/datasets/gsamj462eb.txt. This list, along with the eclipsing binary data from earlier AAVSO publications, is also included in the Lichtenknecker database administrated by the Bundesdeutsche Arbeitsgemeinschaft für Veränderliche Sterne e. V. (BAV) at: http://www.bav-astro.de/LkDB/index.php?lang=en. These observations were reduced by the observers or the writer using the method of Kwee and van Woerden (1956). The standard error is included when available. Column F indicates the filter used. A "C" indicates a clear filter.

The linear elements in the *General Catalogue of Variable Stars* (GCVS; Kholopov *et al.* 1985) were used to compute the O–C values for most stars. For a few exceptions where the GCVS elements are missing or are in significant error, light elements from another source are used: CD Cam (Baldwin and Samolyk 2007), AC CMi (Samolyk 2008), CW Cas (Samolyk 1992a), DV Cep (Frank and Lichtenknecker 1987), Z Dra (Danielkiewicz-Krośniak and Kurpińska-Winiarska 1996), DF Hya (Samolyk 1992b), DK Hya (Samolyk 1990), and GU Ori (Samolyk 1985).

The light elements used for FS Aqr, IR Cnc, TY CMi, AP CMi, BH CMi, CZ CMi, V728 Her, V899 Her, V1033 Her, V1034 Her, WZ Leo, V351 Peg, DS Psc, DZ Psc, GR Psc, V1123 Tau, V1128 Tau, BD Vir, HT Vir, and MS Vir are from (Kreiner 2018).

The light elements used for DD Aqr, V1542 Aql, XY Boo, GH Boo, GM Boo, IK Boo, CW CMi, CX CMi, BD CrB, V1065 Her, V1092 Her, V1097 Her, V470 Hya, V474 Hya, XX Leo, CE Leo, GU Leo, GV Leo, HI Leo, V2610 Oph, V1853 Ori, V2783 Ori, KV Peg, VZ Psc, ET Psc, V1370 Tau, QT UMa, IR Vir, and NN Vir are from (Paschke 2014).

The light elements used for V359 Aur, V337 Gem, and HO Psc are from (Nelson 2014).

The light elements used for V380 Gem, V388 Gem, EU Hya, V409 Hya, and V391 Vir are from the AAVSO VSX site (Watson *et al.* 2014). O–C values listed in this paper can be directly compared with values published in the AAVSO EB monographs.

References

Baldwin, M. E., and Samolyk, G. 2007, *Observed Minima Timings of Eclipsing Binaries No. 12*, AAVSO, Cambridge, MA.

Danielkiewicz-Krośniak, E., and Kurpińska-Winiarska, M., eds. 1996, *Rocznik Astron.* (SAC 68), **68**, 1.

Frank, P., and Lichtenknecker, D. 1987, *BAV Mitt.*, No. 47, 1.

Kholopov, P. N., *et al.* 1985, *General Catalogue of Variable Stars*, 4th ed., Moscow.

Kreiner, J. M. 2004, *Acta Astron.*, **54**, 207 (http://www.as.up.krakow.pl/ephem/).

Kwee, K. K., and van Woerden, H. 1956, *Bull. Astron. Inst. Netherlands*, **12**, 327.

Nelson, R. 2014, Eclipsing Binary O–C Files (http://www.aavso.org/bob-nelsons-o-c-files).

Paschke, A. 2014, "O–C Gateway" (http://var.astro.cz/ocgate/).

Samolyk, G. 1985, *J. Amer. Assoc. Var. Star Obs.*, **14**, 12.

Samolyk, G. 1990, *J. Amer. Assoc. Var. Star Obs.*, **19**, 5.

Samolyk, G. 1992a, *J. Amer. Assoc. Var. Star Obs.*, **21**, 34.

Samolyk, G. 1992b, *J. Amer. Assoc. Var. Star Obs.*, **21**, 111.

Samolyk, G. 2008, *J. Amer. Assoc. Var. Star Obs.*, **36**, 171.

Watson, C., Henden, A. A., and Price, C. A. 2014, AAVSO International Variable Star Index VSX (Watson+, 2006–2014; http://www.aavso.org/vsx).

Table 1. Recent times of minima of stars in the AAVSO eclipsing binary program.

Star	JD (min) Hel. 2400000+	Cycle	O–C (day)	F	Observer	Error (day)	Star	JD (min) Hel. 2400000+	Cycle	O–C (day)	F	Observer	Error (day)
TW And	58343.7882	4687	–0.0639	V	G. Samolyk	0.0001	BI CVn	58304.7281	36286.5	–0.3434	V	S. Cook	0.0003
WZ And	58327.8114	25092	0.0809	V	G. Samolyk	0.0002	R CMa	58170.6803	12220	0.1264	R	G. Samolyk	0.0002
AB And	58124.6052	66332	–0.0442	V	S. Cook	0.0009	RT CMa	58161.7611	24376	–0.7722	V	G. Samolyk	0.0001
AB And	58303.8253	66872	–0.0458	V	G. Samolyk	0.0001	TU CMa	58165.7289	27654	–0.0107	V	G. Samolyk	0.0001
AB And	58333.8615	66962.5	–0.0459	V	R. Sabo	0.0001	TZ CMa	58162.7573	16238	–0.1738	V	G. Samolyk	0.0001
AD And	58337.7607	19606	–0.0458	V	G. Samolyk	0.0003	TZ CMa	58164.6165	16239	–0.2261	V	G. Samolyk	0.0001
BD And	58131.6036	50052	0.0177	V	S. Cook	0.0007	TY CMi	57815.7079	4091	–0.0096	C	G. Frey	0.0001
DS And	58327.8453	21954.5	0.0049	V	G. Samolyk	0.0002	XZ CMi	58152.7161	27139	0.0034	V	K. Menzies	0.0001
QR And	58340.7813	34130	0.1540	V	K. Menzies	0.0003	XZ CMi	58181.6573	27189	0.0041	V	G. Samolyk	0.0001
RY Aqr	58360.3891	8917	–0.1436	V	T. Arranz	0.0001	YY CMi	58212.6364	27595	0.0158	V	S. Cook	0.0008
CX Aqr	58014.7140	38780	0.0142	C	G. Frey	0.0001	AC CMi	57785.6371	6696	0.0038	C	G. Frey	0.0001
CZ Aqr	58361.7546	17375	–0.0651	V	G. Samolyk	0.0001	AC CMi	58203.6374	7178	0.0056	V	G. Samolyk	0.0001
DD Aqr	58015.7211	14132	0.0007	C	G. Frey	0.0001	AK CMi	58226.3925	26727	–0.0220	V	T. Arranz	0.0001
EX Aqr	58054.6926	6245	0.0182	C	G. Frey	0.0002	AP CMi	57789.6888	2445	–0.0325	C	G. Frey	0.0004
FS Aqr	58016.6720	21051	–0.0012	C	G. Frey	0.0001	BH CMi	57784.7066	9449	0.0022	C	G. Frey	0.0002
KO Aql	58361.6238	5752	0.1070	V	G. Samolyk	0.0001	CW CMi	57799.6922	17794.5	–0.0401	C	G. Frey	0.0002
OO Aql	58306.8405	38859.5	0.0714	V	G. Samolyk	0.0001	CX CMi	57813.6770	5225	0.0233	C	G. Frey	0.0002
OO Aql	58349.4106	38943.5	0.0712	V	T. Arranz	0.0001	CZ CMi	57771.6755	12363	–0.0120	C	G. Frey	0.0002
OO Aql	58350.4244	38945.5	0.0715	V	T. Arranz	0.0001	TY Cap	58323.4376	9505	0.0953	V	T. Arranz	0.0002
V342 Aql	58327.7582	5606	–0.1073	V	G. Samolyk	0.0002	RZ Cas	58341.7753	12668	0.0800	V	G. Samolyk	0.0001
V346 Aql	58343.4355	14846	–0.0136	V	T. Arranz	0.0001	TV Cas	58343.7098	7581	–0.0308	V	G. Samolyk	0.0002
V417 Aql	58018.7106	40532.5	0.0610	C	G. Frey	0.0001	CW Cas	58341.6617	52403.5	–0.1182	V	G. Samolyk	0.0002
V609 Aql	58019.6905	35972	–0.0707	C	G. Frey	0.0002	IR Cas	58326.8214	23451	0.0141	V	G. Samolyk	0.0001
V724 Aql	58039.6972	5848	–0.0168	C	G. Frey	0.0002	IS Cas	58306.8265	16036	0.0707	V	G. Samolyk	0.0001
V1542 Aql	58020.7212	14151	0.0140	C	G. Frey	0.0002	OR Cas	58306.8290	11316	–0.0325	V	G. Samolyk	0.0001
RX Ari	58103.6117	19138	0.0597	C	G. Frey	0.0004	OX Cas	58148.6421	6780.5	0.0184	V	S. Cook	0.0009
SS Ari	58154.5701	47110.5	–0.3864	V	G. Samolyk	0.0002	PV Cas	58330.7305	10342	–0.0332	V	G. Samolyk	0.0002
SX Aur	58158.6693	14872	0.0211	SG	G. Conrad	0.0002	V375 Cas	58316.8609	16068	0.2668	V	G. Samolyk	0.0002
TT Aur	58143.6822	27688.5	–0.0072	SG	G. Conrad	0.0002	U Cep	58228.6497	5490	0.2158	V	G. Samolyk	0.0002
AP Aur	58191.6212	27658.5	1.6824	V	G. Samolyk	0.0001	SU Cep	58302.6749	35475	0.0059	V	G. Samolyk	0.0001
EP Aur	58176.5530	53920	0.0196	V	K. Menzies	0.0001	SU Cep	58335.5766	35511.5	0.0064	V	T. Arranz	0.0001
EP Aur	58199.6011	53959	0.0184	V	G. Samolyk	0.0001	SU Cep	58341.4344	35518	0.0051	V	T. Arranz	0.0001
HP Aur	58162.6346	10828	0.0658	V	K. Menzies	0.0001	VW Cep	58154.4408	50293	–0.2485	TG	A. Nemes	0.0005
V459 Aur	58151.7053	1044	0.0068	V	S. Cook	0.0008	VW Cep	58168.3527	50343	–0.2523	TG	A. Nemes	0.0005
TU Boo	58231.7121	77602.5	–0.1583	V	G. Samolyk	0.0001	WW Cep	58302.8357	21696	0.3551	V	G. Samolyk	0.0001
TY Boo	58192.8325	74767.5	0.0668	V	G. Samolyk	0.0001	WZ Cep	58326.7799	72551.5	–0.1894	V	G. Samolyk	0.0002
TY Boo	58204.8834	74805.5	0.0661	V	K. Menzies	0.0001	WZ Cep	58361.6332	72635	–0.1929	V	G. Samolyk	0.0002
TY Boo	58238.6604	74912	0.0669	V	G. Samolyk	0.0001	XX Cep	58302.8260	5760	0.0226	V	G. Samolyk	0.0001
TY Boo	58254.6756	74962.5	0.0661	V	G. Samolyk	0.0001	DK Cep	58356.5613	25120	0.0295	V	T. Arranz	0.0001
TY Boo	58255.4689	74965	0.0666	V	T. Arranz	0.0001	DL Cep	58307.6896	14941	0.0650	V	G. Samolyk	0.0001
TY Boo	58255.6270	74965.5	0.0661	V	T. Arranz	0.0001	DL Cep	58356.6037	14971	0.0646	V	T. Arranz	0.0001
TY Boo	58297.6489	75098	0.0659	V	G. Samolyk	0.0001	DV Cep	58237.8481	9875	–0.0060	V	G. Samolyk	0.0002
TY Boo	58305.7315	75123.5	0.0613	V	S. Cook	0.0008	DV Cep	58308.7296	9936	–0.0050	V	G. Samolyk	0.0001
TZ Boo	58195.8711	62467.5	0.0621	V	G. Samolyk	0.0001	EG Cep	58216.8696	28685	0.0099	V	G. Samolyk	0.0002
TZ Boo	58237.6225	62608	0.0622	V	G. Samolyk	0.0001	EG Cep	58299.6524	28837	0.0102	V	G. Samolyk	0.0001
TZ Boo	58237.7716	62608.5	0.0627	V	G. Samolyk	0.0001	EK Cep	58046.6729	4301	0.0129	V	G. Samolyk	0.0001
TZ Boo	58254.4122	62664.5	0.0623	V	T. Arranz	0.0001	EK Cep	58077.6664	4308	0.0119	V	S. Cook	0.0005
TZ Boo	58254.5600	62665	0.0615	V	T. Arranz	0.0001	TT Cet	58067.6978	52519	–0.0816	C	G. Frey	0.0001
TZ Boo	58301.6603	62823.5	0.0616	V	G. Samolyk	0.0002	TT Cet	58136.7062	52661	–0.0790	V	S. Cook	0.0004
UW Boo	58204.7923	15726	–0.0027	V	K. Menzies	0.0001	VV Cet	58056.7051	51165	0.1377	C	G. Frey	0.0001
VW Boo	58187.9018	78915.5	–0.2764	V	G. Samolyk	0.0002	RW Com	58191.6113	76551.5	0.0103	V	G. Samolyk	0.0003
VW Boo	58287.6841	79207	–0.2819	V	S. Cook	0.0005	RW Com	58191.7298	76552	0.0102	V	G. Samolyk	0.0001
VW Boo	58306.6849	79262.5	–0.2802	V	G. Samolyk	0.0001	RW Com	58214.6326	76648.5	0.0091	V	K. Menzies	0.0001
XY Boo	57876.7248	48366	0.0144	C	G. Frey	0.0002	RZ Com	58243.6451	69145.5	0.0559	V	G. Samolyk	0.0001
AC Boo	58290.7122	92256.5	0.3755	V	S. Cook	0.0008	RZ Com	58254.6466	69178	0.0560	V	N. Simmons	0.0001
AD Boo	58307.6396	16312	0.0360	V	G. Samolyk	0.0003	SS Com	58246.6139	80532.5	0.9402	V	G. Samolyk	0.0002
ET Boo	58274.7352	5104	–0.0111	V	S. Cook	0.0004	CC Com	58134.8703	84288.5	–0.0282	V	B. Harris	0.0001
GH Boo	57878.7036	10036	–0.0021	C	G. Frey	0.0002	CC Com	58152.8560	84370	–0.0284	V	K. Menzies	0.0001
GM Boo	57875.7457	16267	0.0229	C	G. Frey	0.0001	CC Com	58199.5312	84581.5	–0.0284	V	T. Arranz	0.0001
IK Boo	57862.6917	14780	–0.0183	C	G. Frey	0.0002	CC Com	58199.6416	84582	–0.0283	V	T. Arranz	0.0001
SV Cam	58238.6724	26378	0.0581	V	G. Samolyk	0.0002	CC Com	58228.4407	84712.5	–0.0288	V	T. Arranz	0.0001
AL Cam	58086.9379	23846	–0.0222	V	G. Samolyk	0.0001	CC Com	58231.4204	84726	–0.0284	V	T. Arranz	0.0001
AL Cam	58243.6795	23964	–0.0239	V	G. Samolyk	0.0001	CC Com	58253.7086	84827	–0.0295	V	S. Cook	0.0002
CD Cam	58195.6558	7109.5	–0.0108	V	G. Samolyk	0.0002	U CrB	58191.7878	12005	0.1390	C	G. Samolyk	0.0001
WY Cnc	58213.4692	38416	–0.0451	V	T. Arranz	0.0001	U CrB	58288.4517	12033	0.1413	V	T. Arranz	0.0001
IR Cnc	58162.7375	7889	–0.0121	V	K. Menzies	0.0004	RW CrB	58216.8436	24043	0.0041	V	G. Samolyk	0.0002

Table continued on following pages

Table 1. Recent times of minima of stars in the AAVSO eclipsing binary program, cont.

Star	JD (min) Hel. 2400000+	Cycle	O–C (day)	F	Observer	Error (day)	Star	JD (min) Hel. 2400000+	Cycle	O–C (day)	F	Observer	Error (day)
RW CrB	58257.5229	24099	0.0044	V	T. Arranz	0.0001	RZ Dra	58329.5722	25690	0.0688	V	T. Arranz	0.0001
TW CrB	58238.6979	34541	0.0582	V	G. Samolyk	0.0001	TW Dra	58226.6241	5020	–0.0428	V	T. Arranz	0.0001
BD CrB	58213.6218	19599	0.0176	V	T. Arranz	0.0004	UZ Dra	58238.8027	5111	0.0031	V	G. Samolyk	0.0002
BD CrB	58251.4533	19705	0.0207	V	T. Arranz	0.0004	AI Dra	58228.8958	12460	0.0389	V	G. Samolyk	0.0002
W Crv	58192.8094	47786.5	0.0188	V	G. Samolyk	0.0001	AI Dra	58263.6603	12489	0.0378	V	G. Samolyk	0.0002
RV Crv	58246.7056	23041	–0.1140	V	S. Cook	0.0004	BH Dra	58195.8149	10002	–0.0035	V	G. Samolyk	0.0001
RV Crv	58246.7193	23041	–0.1003	V	G. Samolyk	0.0003	SX Gem	58204.5390	28659	–0.0589	V	K. Menzies	0.0001
RV Crv	58255.6865	23053	–0.1002	V	G. Samolyk	0.0001	AF Gem	58171.4849	24937	–0.0709	V	T. Arranz	0.0001
SX Crv	58246.7414	54416	–0.9203	V	G. Samolyk	0.0002	AF Gem	58192.6252	24954	–0.0701	V	G. Samolyk	0.0001
SX Crv	58255.6067	54444	–0.9209	V	G. Samolyk	0.0003	EG Gem	57772.6495	23895	0.3127	C	G. Frey	0.0001
V Crt	58166.8686	23887	0.0000	V	G. Samolyk	0.0002	V337 Gem	57825.7226	2253.5	0.1454	C	G. Frey	0.0005
V Crt	58242.6893	23995	0.0008	V	S. Cook	0.0003	V380 Gem	58176.6355	19497	0.0217	V	K. Menzies	0.0001
V Crt	58254.6227	24012	–0.0004	V	G. Samolyk	0.0002	V388 Gem	57788.7514	10298	0.0120	C	G. Frey	0.0001
SW Cyg	58243.8402	3581	–0.3703	V	G. Samolyk	0.0001	SZ Her	58200.8781	19969	–0.0316	V	G. Samolyk	0.0001
SW Cyg	58344.4460	3603	–0.3735	V	T. Arranz	0.0002	SZ Her	58287.5960	20075	–0.0321	V	T. Arranz	0.0001
WW Cyg	58320.5269	5408	0.1461	V	T. Arranz	0.0001	SZ Her	58327.6824	20124	–0.0326	V	G. Samolyk	0.0001
ZZ Cyg	58301.8061	21160	–0.0748	V	G. Samolyk	0.0001	TT Her	58243.8469	20007	0.0452	V	G. Samolyk	0.0002
ZZ Cyg	58324.4374	21196	–0.0737	V	T. Arranz	0.0001	TT Her	58310.4281	20080	0.0449	V	L. Corp	0.0002
ZZ Cyg	58363.4111	21258	–0.0742	V	T. Arranz	0.0001	TU Her	58311.5729	6286	–0.2542	V	T. Arranz	0.0001
AE Cyg	58314.7609	14165	–0.0045	V	G. Samolyk	0.0001	TU Her	58327.4408	6293	–0.2553	V	T. Arranz	0.0001
BR Cyg	58297.7932	12576	0.0010	V	G. Samolyk	0.0001	UX Her	58275.7323	12011	0.1416	V	G. Samolyk	0.0001
BR Cyg	58344.4321	12611	0.0002	V	T. Arranz	0.0001	UX Her	58317.5526	12038	0.1430	V	T. Arranz	0.0001
CG Cyg	58322.8236	29942	0.0777	V	G. Samolyk	0.0001	UX Her	58331.4925	12047	0.1433	V	T. Arranz	0.0001
CG Cyg	58340.4957	29970	0.0778	V	T. Arranz	0.0001	AK Her	57914.7301	37313	0.0193	C	G. Frey	0.0002
DK Cyg	58333.5404	43200	0.1248	V	T. Arranz	0.0001	AK Her	58308.4310	38247	0.0187	V	L. Corp	0.0001
DK Cyg	58347.4268	43229.5	0.1259	V	T. Arranz	0.0001	AK Her	58312.4376	38256.5	0.0208	V	L. Corp	0.0001
DK Cyg	58347.6629	43230	0.1266	V	T. Arranz	0.0002	AK Her	58322.7610	38281	0.0170	V	S. Cook	0.0003
KR Cyg	58265.8611	34502	0.0241	V	G. Samolyk	0.0001	CC Her	57901.7202	10515	0.3072	C	G. Frey	0.0001
KV Cyg	58065.6276	10073	0.0561	V	G. Samolyk	0.0002	CC Her	58246.7969	10714	0.3168	V	G. Samolyk	0.0001
KV Cyg	58326.8159	10165	0.0570	V	G. Samolyk	0.0003	CC Her	58300.5519	10745	0.3176	V	T. Arranz	0.0001
KV Cyg	58332.4948	10167	0.0579	V	T. Arranz	0.0001	CT Her	58210.8871	8782	0.0116	V	G. Samolyk	0.0002
KV Cyg	58349.5299	10173	0.0590	V	T. Arranz	0.0001	HS Her	58323.7399	8039	–0.0341	V	S. Cook	0.0005
V346 Cyg	58343.7096	8259	0.1936	V	G. Samolyk	0.0002	LT Her	58361.5859	16241	–0.1613	V	G. Samolyk	0.0002
V387 Cyg	58301.7063	47325	0.0207	V	G. Samolyk	0.0001	V728 Her	58228.8001	12155	0.0172	V	G. Samolyk	0.0001
V387 Cyg	58333.7367	47375	0.0213	V	R. Sabo	0.0001	V728 Her	58271.6887	12246	0.0180	V	N. Simmons	0.0001
V388 Cyg	58299.8300	19029	–0.1262	V	G. Samolyk	0.0001	V899 Her	57910.7414	12846	–0.0072	C	G. Frey	0.0003
V388 Cyg	58319.5872	19052	–0.1268	V	T. Arranz	0.0001	V1033 Her	57924.7447	18200	–0.0029	C	G. Frey	0.0002
V388 Cyg	58343.6417	19080	–0.1254	V	G. Samolyk	0.0001	V1034 Her	57911.6953	6637	–0.0037	C	G. Frey	0.0002
V401 Cyg	58254.8101	24745	0.0952	V	G. Samolyk	0.0001	V1065 Her	57923.7363	16658	–0.0105	C	G. Frey	0.0004
V401 Cyg	58306.6717	24834	0.0946	V	G. Samolyk	0.0002	V1092 Her	57915.7188	14294	–0.0210	C	G. Frey	0.0004
V456 Cyg	58299.7861	15023	0.0527	V	G. Samolyk	0.0002	V1097 Her	57918.7120	15118	0.0050	C	G. Frey	0.0001
V456 Cyg	58326.5213	15053	0.0521	V	T. Arranz	0.0001	WY Hya	58217.7129	24646	0.0412	V	S. Cook	0.0009
V466 Cyg	58314.5659	21228	0.0077	V	T. Arranz	0.0001	AV Hya	58154.7670	31433	–0.1161	V	G. Samolyk	0.0001
V466 Cyg	58316.6531	21229.5	0.0075	V	G. Samolyk	0.0001	AV Hya	58232.6740	31547	–0.1174	V	S. Cook	0.0008
V466 Cyg	58330.5687	21239.5	0.0075	V	T. Arranz	0.0001	AV Hya	58235.4077	31551	–0.1173	V	T. Arranz	0.0001
V466 Cyg	58335.4389	21243	0.0072	V	T. Arranz	0.0001	DF Hya	58162.5793	46505	0.0070	V	G. Samolyk	0.0001
V477 Cyg	58275.8628	6002	–0.0387	V	G. Samolyk	0.0001	DF Hya	58195.6403	46605	0.0074	V	G. Samolyk	0.0002
V477 Cyg	58323.5074	6022.5	–0.5074	V	T. Arranz	0.0004	DF Hya	58199.4419	46616.5	0.0071	V	T. Arranz	0.0001
V548 Cyg	58302.6555	7670	0.0226	V	G. Samolyk	0.0001	DF Hya	58200.4336	46619.5	0.0070	V	T. Arranz	0.0001
V704 Cyg	58330.7008	35730	0.0379	V	G. Samolyk	0.0002	DF Hya	58231.6753	46714	0.0065	V	S. Cook	0.0004
V704 Cyg	58345.5389	35756	0.0377	V	T. Arranz	0.0003	DF Hya	58237.6276	46732	0.0079	V	G. Samolyk	0.0001
V836 Cyg	58342.5544	20644	0.0226	V	T. Arranz	0.0002	DI Hya	58234.7094	43989	–0.0406	V	S. Cook	0.0008
V836 Cyg	58361.5043	20673	0.0236	V	T. Arranz	0.0001	DK Hya	58199.6693	29422	0.0005	V	G. Samolyk	0.0001
V1034 Cyg	58254.7974	15678	0.0142	V	G. Samolyk	0.0002	DK Hya	58205.4095	29433	–0.0005	V	T. Arranz	0.0001
TT Del	58341.8359	4566	–0.1125	V	G. Samolyk	0.0004	DK Hya	58236.7238	29493	–0.0014	V	S. Cook	0.0006
TY Del	58360.7875	12930	0.0718	V	G. Samolyk	0.0001	EU Hya	57826.6412	30455	–0.0337	C	G. Frey	0.0002
YY Del	58337.5816	19390	0.0118	V	T. Arranz	0.0002	V409 Hya	57809.7205	9857	0.0631	C	G. Frey	0.0001
FZ Del	58340.4392	34494	–0.0252	V	T. Arranz	0.0001	V470 Hya	57807.7494	12914	0.0100	C	G. Frey	0.0002
Z Dra	58181.7692	6142	–0.0019	V	G. Samolyk	0.0001	V474 Hya	57800.7375	10630	–0.0146	C	G. Frey	0.0001
Z Dra	58192.6283	6150	–0.0022	B	G. Lubcke	0.0001	SW Lac	58337.7551	40728.5	–0.0738	V	G. Samolyk	0.0003
Z Dra	58192.6285	6150	–0.0021	V	G. Lubcke	0.0001	VX Lac	58301.8377	12139	0.0866	V	G. Samolyk	0.0001
Z Dra	58192.6286	6150	–0.0020	Ic	G. Lubcke	0.0001	AR Lac	58341.7165	8445	–0.0526	V	G. Samolyk	0.0002
RZ Dra	58228.7627	25507	0.0692	V	G. Samolyk	0.0001	CM Lac	58359.5198	19526	–0.0044	V	T. Arranz	0.0001
RZ Dra	58265.6716	25574	0.0695	V	G. Samolyk	0.0001	DG Lac	58361.7733	6278	–0.2342	V	G. Samolyk	0.0002
RZ Dra	58303.6816	25643	0.0692	V	G. Samolyk	0.0001	Y Leo	58203.5484	7572	–0.0669	V	K. Menzies	0.0001
RZ Dra	58318.5560	25670	0.0701	V	T. Arranz	0.0001	UU Leo	58249.3687	7651	0.2151	V	T. Arranz	0.0001

Table continued on following pages

Table 1. Recent times of minima of stars in the AAVSO eclipsing binary program, cont.

Star	JD (min) Hel. 2400000+	Cycle	O–C (day)	F	Observer	Error (day)	Star	JD (min) Hel. 2400000+	Cycle	O–C (day)	F	Observer	Error (day)
UV Leo	57830.7082	32312	0.0425	C	G. Frey	0.0004	VZ Psc	58035.6670	54365.5	−0.0011	C	G. Frey	0.0002
UV Leo	58262.7725	33032	0.0457	V	N, Krumm	0.0002	DS Psc	58040.7109	16177	−0.0032	C	G. Frey	0.0001
VZ Leo	57839.7120	24475	−0.0533	C	G. Frey	0.0007	DZ Psc	58046.6741	15149	0.0160	C	G. Frey	0.0001
VZ Leo	58238.6220	24841	−0.0489	V	G. Samolyk	0.0003	ET Psc	58043.7108	12329	−0.0043	C	G. Frey	0.0001
WZ Leo	57827.7007	3783	−0.0015	C	G. Frey	0.0001	GR Psc	58050.6777	13317	−0.0011	C	G. Frey	0.0001
WZ Leo	58210.7266	4055	−0.0007	V	G. Samolyk	0.0002	HO Psc	58049.6397	2245	0.0005	C	G. Frey	0.0002
XX Leo	57837.7249	9419	−0.0150	C	G. Frey	0.0003	UZ Pup	58197.6931	17090	−0.0122	V	S. Cook	0.0002
XY Leo	58212.7337	46245	0.1820	V	S. Cook	0.0008	AV Pup	58203.6405	48577	0.2348	V	G. Samolyk	0.0001
XY Leo	58216.7046	46259	0.1755	V	G. Samolyk	0.0003	U Sge	58316.5124	12183	0.0157	V	T. Arranz	0.0001
XZ Leo	58216.7184	27046	0.0769	V	G. Samolyk	0.0003	V505 Sgr	58324.7322	11720	−0.1132	V	G. Samolyk	0.0001
AM Leo	58256.7106	43093	0.0142	V	S. Cook	0.0006	V1968 Sgr	58303.7969	36366	−0.0180	V	G. Samolyk	0.0003
CE Leo	57832.7013	33461	−0.0095	C	G. Frey	0.0001	RS Sct	58327.7116	39131	0.0428	V	G. Samolyk	0.0002
GU Leo	57881.7277	15523	0.0042	C	G. Frey	0.0003	AO Ser	58195.9169	27364	−0.0107	V	G. Samolyk	0.0001
GV Leo	57808.7121	18949	−0.0396	C	G. Frey	0.0001	AO Ser	58234.6070	27408	−0.0119	V	T. Arranz	0.0001
HI Leo	57865.7105	16448	0.0146	C	G. Frey	0.0001	AO Ser	58242.5219	27417	−0.0111	V	T. Arranz	0.0001
T LMi	58216.6484	4245	−0.1314	V	G. Samolyk	0.0001	AO Ser	58336.6112	27524	−0.0120	V	K. Menzies	0.0001
Z Lep	58158.7243	30929	−0.1979	V	G. Samolyk	0.0001	CC Ser	58162.9264	40077.5	1.1179	V	G. Samolyk	0.0002
Z Lep	58170.6491	30941	−0.1977	V	G. Samolyk	0.0001	CC Ser	58238.7838	40224.5	1.1224	V	G. Samolyk	0.0002
SS Lib	58299.6580	11922	0.1823	V	G. Samolyk	0.0001	Y Sex	58218.7034	39189	−0.0203	V	S. Cook	0.0011
RY Lyn	58234.6202	10695	−0.0195	V	G. Samolyk	0.0002	Y Sex	58234.6571	39227	−0.0199	V	G. Samolyk	0.0001
UZ Lyr	58322.6716	7737	−0.0468	V	G. Samolyk	0.0001	RZ Tau	58096.6741	49125	0.0875	C	G. Frey	0.0001
UZ Lyr	58343.4752	7748	−0.0472	V	T. Arranz	0.0001	RZ Tau	58151.5433	49257	0.0876	V	G. Samolyk	0.0002
EW Lyr	58238.8389	16287	0.2904	V	G. Samolyk	0.0001	RZ Tau	58195.6062	49363	0.0890	V	G. Samolyk	0.0001
EW Lyr	58289.5074	16313	0.2921	V	T. Arranz	0.0001	AM Tau	58154.6023	6312	−0.0756	V	G. Samolyk	0.0001
EW Lyr	58316.7901	16327	0.2927	V	G. Samolyk	0.0001	EQ Tau	58143.6404	52528	−0.0376	V	S. Cook	0.0001
FL Lyr	58275.8174	9207	−0.0027	V	G. Samolyk	0.0001	HU Tau	58124.6782	8194	0.0366	V	S. Cook	0.0007
FL Lyr	58332.4498	9233	−0.0023	V	T. Arranz	0.0001	V1123 Tau	58175.3331	14190	0.0104	V	L. Corp	0.0003
RU Mon	58151.6832	4577.5	−0.7000	V	G. Samolyk	0.0001	V1128 Tau	57783.6824	17302	−0.0004	C	G. Frey	0.0009
RU Mon	58157.6271	4579	−0.1333	V	G. Samolyk	0.0001	V1370 Tau	58152.6077	22620	0.0022	V	K. Menzies	0.0001
RW Mon	58200.3552	12864	−0.0877	V	T. Arranz	0.0001	W UMa	58199.6325	37268	−0.1080	V	G. Samolyk	0.0002
AT Mon	58154.7150	15548	0.0120	V	G. Samolyk	0.0001	W UMa	58199.8002	37268.5	−0.1071	V	G. Samolyk	0.0001
BO Mon	58160.6498	6585	−0.0163	V	G. Samolyk	0.0001	W UMa	58218.6490	37325	−0.1088	V	G. Conrad	0.0002
BO Mon	58227.4067	6615	−0.0160	V	T. Arranz	0.0001	TY UMa	58137.8465	52476.5	0.4052	V	B. Harris	0.0001
U Oph	58215.9048	8227	−0.0085	V	N. Simmons	0.0002	TY UMa	58151.8522	52516	0.4066	V	G. Samolyk	0.0002
SX Oph	58305.6934	12071	−0.0038	V	G. Samolyk	0.0002	TY UMa	58174.5428	52580	0.4067	V	T. Arranz	0.0001
V508 Oph	58253.5756	38200	−0.0267	V	T. Arranz	0.0001	TY UMa	58195.6386	52639.5	0.4075	V	N. Simmons	0.0001
V508 Oph	58290.4694	38307	−0.0257	V	L. Corp	0.0002	TY UMa	58195.6390	52639.5	0.4079	V	G. Lubcke	0.0002
V508 Oph	58306.6738	38354	−0.0265	V	G. Samolyk	0.0001	TY UMa	58195.6392	52639.5	0.4080	Ic	G. Lubcke	0.0001
V839 Oph	58288.4977	43618.5	0.3224	V	T. Arranz	0.0002	TY UMa	58195.6392	52639.5	0.4080	B	G. Lubcke	0.0003
V839 Oph	58322.6506	43702	0.3242	V	G. Samolyk	0.0001	TY UMa	58237.6536	52758	0.4096	V	G. Samolyk	0.0002
V839 Oph	58325.5135	43709	0.3242	V	T. Arranz	0.0001	TY UMa	58258.7501	52817.5	0.4111	V	N, Krumm	0.0001
V1010 Oph	58255.8441	29207	−0.1979	V	G. Samolyk	0.0001	TY UMa	58280.7328	52879.5	0.4124	V	S. Cook	0.0003
V1010 Oph	58316.6942	29299	−0.1990	V	G. Samolyk	0.0001	UX UMa	58154.8923	105364	−0.0008	V	G. Samolyk	0.0001
V2610 Oph	58292.4598	13886	−0.0358	V	L. Corp	0.0003	UX UMa	58192.6529	105556	−0.0011	V	G. Samolyk	0.0001
EQ Ori	58174.3239	15312	−0.0439	V	T. Arranz	0.0001	UX UMa	58193.6362	105561	−0.0011	V	G. Lubcke	0.0004
FL Ori	58161.6287	8262	0.0423	V	G. Samolyk	0.0001	UX UMa	58228.6435	105739	−0.0014	V	G. Samolyk	0.0001
FZ Ori	58173.3537	35373.5	−0.0306	V	T. Arranz	0.0001	UX UMa	58246.7377	105831	−0.0009	V	K. Menzies	0.0001
GU Ori	58151.6338	32042.5	−0.0651	V	G. Samolyk	0.0002	VV UMa	58175.7277	17982	−0.0760	V	G. Lubcke	0.0001
GU Ori	58174.4621	32091	−0.0649	V	T. Arranz	0.0001	VV UMa	58175.7277	17982	−0.0759	B	G. Lubcke	0.0002
GU Ori	58175.4032	32093	−0.0651	V	T. Arranz	0.0001	VV UMa	58175.7277	17982	−0.0758	Ic	G. Lubcke	0.0002
V1853 Ori	57781.6782	9700	0.0004	C	G. Frey	0.0002	VV UMa	58176.7555	17983.5	−0.0793	B	G. Lubcke	0.0032
V2783 Ori	57782.7478	1147	0.0098	C	G. Frey	0.0001	VV UMa	58176.7581	17983.5	−0.0767	Ic	G. Lubcke	0.0003
U Peg	58307.8509	58157.5	−0.1689	V	G. Samolyk	0.0001	VV UMa	58176.7598	17983.5	−0.0749	V	G. Lubcke	0.0007
AQ Peg	58351.5126	3087	0.5797	V	T. Arranz	0.0001	VV UMa	58209.4097	18031	−0.0756	V	T. Arranz	0.0001
BB Peg	58361.7591	40380	−0.0291	V	G. Samolyk	0.0001	XZ UMa	58183.6836	9830	−0.1480	V	S. Cook	0.0003
BX Peg	58319.7607	50369	−0.1317	V	K. Menzies	0.0001	XZ UMa	58216.6868	9857	−0.1474	V	G. Samolyk	0.0001
BX Peg	58341.6337	50447	−0.1315	V	G. Samolyk	0.0001	XZ UMa	58226.4649	9865	−0.1479	V	T. Arranz	0.0001
DI Peg	58017.7402	18012	0.0080	C	G. Frey	0.0001	ZZ UMa	58210.6188	9681	−0.0013	V	G. Samolyk	0.0002
DI Peg	58136.6148	18179	0.0092	V	S. Cook	0.0005	ZZ UMa	58272.6965	9708	−0.0036	V	S. Cook	0.0003
DK Peg	58041.7259	7667	0.1596	C	G. Frey	0.0002	AF UMa	58226.6889	5978	0.6251	V	S. Cook	0.0009
EE Peg	58042.6613	4748	0.0085	C	G. Frey	0.0002	QT UMa	58176.6860	13964	0.0094	V	K. Menzies	0.0001
KV Peg	58025.7141	22419	−0.0227	C	G. Frey	0.0003	W UMi	58195.7957	14411	−0.2083	V	N. Simmons	0.0001
V351 Peg	58021.7245	16049	0.0374	C	G. Frey	0.0002	W UMi	58275.7480	14457	−0.2104	V	G. Samolyk	0.0003
RT Per	58162.5201	29181	0.1111	V	K. Menzies	0.0001	RU UMi	58200.7862	31632	−0.0152	V	G. Samolyk	0.0001
IU Per	58131.7243	14609	0.0086	V	S. Cook	0.0007	AG Vir	57858.6938	19336	−0.0157	C	G. Frey	0.0002
KW Per	58155.6616	16914	0.0176	V	S. Cook	0.0003	AG Vir	58243.6411	19935	−0.0162	V	G. Samolyk	0.0002

Table continued on next page

Table 1. Recent times of minima of stars in the AAVSO eclipsing binary program, cont.

Star	JD (min) Hel. 2400000 +	Cycle	O–C (day)	F	Observer	Error (day)	Star	JD (min) Hel. 2400000 +	Cycle	O–C (day)	F	Observer	Error (day)
AH Vir	58230.4192	30466.5	0.2957	V	L. Corp	0.0001	IR Vir	57864.6857	21763.5	–0.0100	C	G. Frey	0.0001
AH Vir	58275.6544	30577.5	0.2960	V	G. Samolyk	0.0001	MS Vir	57912.6932	17324	–0.0021	C	G. Frey	0.0003
AK Vir	58249.5036	13131	–0.0411	V	T. Arranz	0.0001	NN Vir	57877.7568	19508	0.0067	C	G. Frey	0.0002
AW Vir	57871.7013	36297	0.0290	C	G. Frey	0.0003	V391 Vir	57874.6918	18670	0.0041	C	G. Frey	0.0001
AW Vir	58168.8835	37136.5	0.0308	V	G. Samolyk	0.0001	Z Vul	58305.5297	6256	–0.0151	V	T. Arranz	0.0001
AW Vir	58265.7007	37410	0.0298	V	S. Cook	0.0005	AW Vul	58342.6867	14951	–0.0333	V	G. Samolyk	0.0002
AZ Vir	57890.7405	39793	–0.0232	C	G. Frey	0.0001	AX Vul	58336.7517	6659	–0.0385	V	K. Menzies	0.0001
AZ Vir	58216 8054	40725.5	–0.0210	V	G. Samolyk	0.0001	AY Vul	58006.6773	6351	–0.1493	V	G. Samolyk	0.0003
BD Vir	58270.7149	6173	0.1830	V	S. Cook	0.0006	AY Vul	58305.8074	6475	–0.1626	V	G. Samolyk	0.0002
BF Vir	57895.7265	18460	0.1203	C	G. Frey	0.0001	BE Vul	58307.6524	11724	0.1075	V	G. Samolyk	0.0001
BH Vir	57896.7093	17954	–0.0126	C	G. Frey	0.0001	BE Vul	58363.5268	11760	0.1084	V	T. Arranz	0.0001
BH Vir	58158.9246	18275	–0.0131	V	G. Samolyk	0.0001	BO Vul	58337.7232	11485	–0.0133	V	G. Samolyk	0.0001
BH Vir	58214.4720	18343	–0.0129	V	T. Arranz	0.0001	BS Vul	58343.6562	31666	–0.0344	V	G. Samolyk	0.0001
BH Vir	58271.6536	18413	–0.0124	V	G. Samolyk	0.0001	BT Vul	58327.7528	20089	0.0060	V	R. Sabo	0.0002
DL Vir	58269.6930	14803	0.1176	V	S. Cook	0.0008	BU Vul	58263.8411	43463	0.0153	V	G. Samolyk	0.0001
HT Vir	57902.7294	13252	0.0000	C	G. Frey	0.0001	CD Vul	58307.8016	17564	–0.0006	V	G. Samolyk	0.0001
HT Vir	58295.7238	14216	–0.0018	V	S. Cook	0.0008							

Abstracts of Papers Presented at the Joint Meeting of the British Astronomical Association, Variable Star Section and the American Association of Variable Star Observers (AAVSO 107th Spring Meeting), Held in Warwick, United Kingdom, July 7–8, 2018

The HOYS-CAPS Citizen Science Project

Dirk Froebrich
School of Physical Sciences, University of Kent, Canterbury, Kent CT2 7NZ, United Kingdom; D.Froebrich@kent.ac.uk

Abstract   The talk will introduce the science goals of the HOYS-CAPS citizen science project and explain how to participate. We will also show some of the initial results.

Recent Activity of SU Aurigae

Michael Poxon
9 Rosebery Road, Great Plumstead, Norfolk NR13 5EA, United Kingdom; mike@starman.co.uk

Abstract   Observations of the recent anomalous behavior of the T Tauri star SU Aur are used in conjunction with previous studies to better understand the system.

Evidence for Starspots on T Tauri Stars

Andrew Wilson
12 Barnards Close, Yatton, Bristol BS49 4HZ, United Kingdom; barnards.star12@gmail.com

Abstract   Observational Color-Magnitude Diagrams (CMD) of young star clusters show a spread that is indicative of a spread in age. However, it could be that the stars formed at around the same time but a physical property of the stars is at least partially responsible for the spread. One such property is magnetic field of the Young Stellar Object (YSO). A strong magnetic field would inhibit convection, slowing contraction of the YSO towards the main sequence and thus causing a spread in the CMD. Starspots are a good indicator of stellar magnetic field. Spectra of T Tauri stars in the Orion Nebula Cluster and the σ Ori Cluster are being analyzed to discover if they show the presence of a large surface covering by starspots. This work is being undertaken as part of Andrew's Ph.D. project at the University of Exeter under the supervision of Professor Tim Naylor.

The Discovery of TT Crateris

John Toone
17 Ashdale Road, Cressage, Shrewsbury SY5 6DT, United Kingdom; enootnhoj@btinternet.com

Abstract   The discovery of TT Cra was a remarkable achievement by an amateur astronomer. Visual discoveries of dwarf novae are rare and this one at the time of the Comet Halley apparition in 1986 was much fainter than the others that were found by professional astronomers during the period 1855–1904. This presentation explains the circumstances of the discovery and the follow-up efforts by amateur astronomers to obtain its official recognition in 1989.

AR Scorpii: a Remarkable Highly Variable Star Discovered by Amateur Astronomers

Thomas Marsh
Department of Physics, University of Warwick, Coventry CV4 7AL, United Kingdom; t.r.marsh@warwick.ac.uk

Abstract   In May 2015, a group of amateur astronomers contacted Boris Gaensicke at Warwick regarding a puzzling star that they had been observing. This star, AR Sco, has turned out to be one of the most remarkable objects in the sky, unique for astonishingly strong pulsations every two minutes, and for radiating power across the electromagnetic spectrum, from radio to X-ray wavelengths. I will describe what we think AR Sco is, how we arrived at this picture, and the extremely puzzling problems that it continues to pose.

Long Term Orbital Behavior of Eclipsing SW Sextantis Stars

David Boyd
5 Silver Lane, West Challow, Wantage OX12 9TX, United Kingdom; davidboyd@orion.me.uk

Abstract   In 2006, encouraged by Boris Gaensicke, I began a long-term project to investigate the orbital behavior of the 18 brightest eclipsing SW Sex stars. These are novalike CVs in which the high rate of mass transfer between the main sequence secondary star and the white dwarf primary via an accretion disc maintains the system in a persistent bright state. The initial aims of the project were to establish accurate ephemerides for these stars and to check if any of them deviated from a linear ephemeris. At the 100th Spring Meeting of the AAVSO in Boston in May 2011 I presented the results of the first five years of the project, which combined new measurements of eclipse times with previously published observations. At that time, the majority of the stars appeared to be behaving consistently with linear ephemerides. However, five stars indicated possible cyclical variation in their orbital periods and three more were clearly not following linear ephemerides. I now have a further seven years of eclipse observations on these stars and it is time to revisit these earlier conclusions. It seems that linear ephemerides are no longer the most common option. Something is happening to upset the regular orbital behavior in several of these systems.

Seven Years on the ROAD (Remote Observatory Atacama Desert)

Franz-Josef Hambsch
Oude Bleken 12, Mol 2400, Belgium; hambsch@telenet.be

Abstract   After several tries at different places to set up a remote observatory, the ultimate destination has been found in San Pedro de Atacama at Alain Maury's place called Spaceobs. Since its start on August, 1, 2011, the Remote Observatory Atacama Desert (ROAD) has produced tons of data due to the exceptional weather conditions in the Atacama dessert. The hardware and software which is used is mostly off the shelf. A 40-cm optimized Dall Kirkham (ODK) from Orion Optics, UK, is the workhorse, riding on a DDM85 direct drive mount from ASA (AstroSysteme Austria). The CCD is an ML16803 from FLI equipped with Astrodon UBVRI photometrical filters. Analysis of the images is done with the LESVEPHOTOMETRY program written by Pierre de Ponthièrre, an amateur astronomer from Lesve, Belgium. Further software packages in use are MAXIMDL for image acquisition and CCDCOMMANDER for automatization. From the start the focus was on pro-am collaborations and a few examples will be highlighted during the presentation. Most of the data are shared with VSNET in Kyoto, Japan, the Centre of Backyard Astrophysics (CBA), USA and several professional astronomers. Also most of those data are accessible from the AAVSO International Database (AAVSO user code: HMB). Related publications with co-authorship can be found on ARXIV using in the search box my last name.

Long Term Spectroscopic Monitoring of the Brightest Symbiotic Stars

Francois Teyssier
67 Rue Jacques Daviel, Rouen 76100, France; francoismathieu. teyssier@bbox.fr

Abstract   Symbiotic stars are wide interacting binary systems comprising a cool giant and a hot compact star, mostly a white dwarf, accreting from the giant's wind. Their orbital periods are hundreds of days (for S-type systems containing a normal giant). The accreting WD represents a strong source of ultraviolet radiation that ionizes a fraction of the wind from the giant and produces a rich emission spectrum. They are strongly variable, according to orbital phase and activity, and can produce various types of outbursts. Symbiotic stars are considered as excellent laboratories for studying a variety of astrophysical problems, such as wind from red giants, accretion—eventually throw a disk—thermonuclear outbursts under a wide range of conditions, collimation of stellar wind, formation of jets, etc. About 50 symbiotic stars in the galaxy are bright enough to be studied by amateur spectroscopy with small telescopes ranging from 8 to 24 inches. We have undergone a long-term monitoring program in the visual range of the brightest symbiotics at a resolution from 500 to 15,000. A part of this program is performed in collaboration with or upon the request of professional teams, feeding several publications at least partially (for instance: T CrB, AG Peg, AG Dra).

Gaia: Transforming Stellar Astronomy

Boris Gaensicke
Department of Physics, University of Warwick, Coventry CV4 7AL, United Kingdom; Boris.Gaensicke@warwick.ac.uk

Abstract   The only way to measure the distances to stars is via a geometric parallax, making use of the fact that the Earth orbits the Sun. Over a century of work on ground-based parallaxes was limited in reach to a few 100 pc, at best, and much of our understanding of stellar physics had to be based on proxy distance estimates. On April 25, 2018, the ESA Gaia mission unleashed space-based astrometric data for over 1.3 billion sources, transforming stellar astrophysics over lunch time. I will illustrate the quantum leap in stellar astronomy that these data enable, and will discuss how future large spectroscopic and photometric surveys will augment our understanding of stars both in quality and quantity.

Starting in Spectroscopy

Francois Cochard
Shelyak Instruments, 73 rue de Chartreuse 38420, Le Versoud, France

Abstract   Spectroscopy is more and more present in amateur astronomy, and gives deep physical information on the sky objects (stars, nebulae, novae and supernovae, comets...). We'll see how it works in real life: which equipment is required, the optical principles, how to run an observation. I'll also give you some key advice to successfully start in spectroscopy.

Pushing the Limits Using Commercial Spectrographs

Robin Leadbeater
The Birches Torpenhow, Wigton, Cumbria CA7 1JF, United Kingdom; robin@threehillsobservatory.co.uk

Abstract   Some observations which explore the capabilities of three popular spectrograph designs: 1. Simultaneous multi-band photometry of fast transients using a Star Analyser grating; 2. Confirming and classifying magnitude 17 supernovae using an modified ALPY spectrograph; 3. Sub km/sec precision radial velocity measurement using a LHIRES III spectrograph.

Towards Full Automation of High Resolution Spectroscopy

Andrew Smith
Greenacre, 25 Station Road, Delamere CW8 2HU, United Kingdom; andrew.j.smith1905@btinternet.com

Abstract   Following the successful automation of low resolution spectroscopy with a 300-mm F5.4 Newtonian and a LISA spectrograph I decided to move to medium/high resolution with a 400-mm ODK and homemade fibre-fed spectrographs R ~ 10,000–20,000. This talk discusses the construction of the

Medium Resolution echelle spectrograph (R ~ 10,000) and the work necessary to automate its operation to the point where I can supply it with target information, press "Run" on my PYTHON program, and retreat to the comfort of my arm chair. The R ~ 10,000 echelle spectrograph is intended for accurate radial velocity measurement and to this end is temperature stabilized to better than ± 0.04 degree. It uses a conventional layout with a R2 echelle and a F2 prism as cross disperser. Both the collimator and camera lenses are commercial camera lenses. The route to automation rests on the core capabilities and script-ability of Software Bisque's THE SKY X and the accuracy of the Paramount ME II. However, there are a number of challenges due to the small field of view provided by the Shelyak Instruments Fibre Guide-head at the 2.7-m focal length of the 400-mm ODK and the need to center and maintain the target on a 75-micron hole. The separation of the finding and guiding tasks by using a dichroic beam splitter is central to the solution.

Applying Transformation and Extinction to Magnitude Estimate—How Much Does It Improve Results?

Gordon Myers
5 Inverness Way, Hillsborough, CA 94010; GordonMyers@ hotmail.com

Ken Menzies
318A Potter Road, Framingham, MA 01701; kenmenstar@gmail.com

Abstract   Photometrists regularly ask the question as to whether they should apply transformation and extinction corrections to their magnitude estimates. How much do these corrections improve the accuracy of their reported standard magnitudes? How much effort is involved in making these corrections? We quantify the significance of these corrections based on the characteristics of equipment (e.g., filter, CCD and field of view) and the conditions of the observation (e.g., airmass). Specific examples are presented for both CCD and DSLR systems. We discuss the best practices that one should follow to improve their reported magnitudes and the AAVSO tools (VPHOT, Transform Generator, Transform Applier) that facilitate an easy correction to your results. It is found that magnitude corrections for CCD observers are small but significant for most amateur equipment, and critical for most DSLR observers.

Red Dots Initiative: Science and Opportunities in Finding Planets Around the Nearest Red-Dwarfs

Guillem Anglada Escude
School of Physics and Astronomy, Queen Mary University of London, G. O. Jones Building, 327 Mile End Road, London E1 4NS, United Kingdom; g.anglada@qmul.ac.uk

Abstract   Nearby red dwarf stars are ideal grounds to search for small planets. The Pale Red Dot campaign (2016) consisted in continuously monitoring of Proxima Centauri with the HARPS spectrometer. This campaign was aimed at measuring the motion of the star caused by a planet orbiting it using the Doppler effect. Although this is a mature technique to find planets, we are at the level where stellar activity contaminates the Doppler measurements and it is at the same level of the planetary signals under investigation. For this reason, additional information needs to be collected from the star. In particular, quasi-simultaneous photometric observations to the Doppler measurements are very useful to distinguish certain kinds of spurious signals from true planets. In 2017 we performed a second campaign called Red Dots where three more very nearby red-dwarfs were monitored spectroscopically and photometrically over three months. Many of the photometric observations were contributed by several pro-am astronomers with moderate size telescopes (~0.4-m apertures), which are ideal for this kind of observations. I will review the status of the project, and discuss further opportunities for pro-am astronomers to contribute to this science cause.

How To Find Planets and Black Holes with Microlensing Events

Lukasz Wyrzykowski
Warsaw University Astronomical Observatory, Department of Physics, Al. Ujazdowskie 4, 00-478 Warszawa, Poland; lw@astrouw.edu.pl

Abstract   As shown by gravitational wave detections, galaxies harbor an unknown population of black holes at high masses. In our Galaxy, dark objects like black holes or planets can be found and studied solely via gravitational microlensing, when a distant source star gets magnified by the space-time curvature caused by the lensing object. In order to measure the mass of the lens, hence to recognize a black hole or a planet, it is necessary to combine highly sampled photometry from the ground with high accuracy astrometric data from Gaia. Well-coordinated observing efforts, as in case of Gaia16aye binary microlensing event, will lead to full characterization and discovery of a population of planets and black holes in the spiral arms of the Milky Way.

Short Period Eclipsing sdB Binaries and the Claims for Circumbinary Objects

George Faillace
D. Pulley
D. Smith
A. Watkins
S. von Harrach
address correspondence to: G. Faillace, Elmore Goring Road, Woodcote, Oxfordshire RG8 0QE, United Kingdom; gfaillace2@aol.com

Abstract   It is well known that two orbiting objects do so around a common center of gravity, or barycenter. What is less well appreciated is that this forms the basis of a powerful astrophysical binary star research tool of which amateurs can make use of as much as their professional colleagues. Our group used this technique to investigate if seemingly periodic variations in the position of the barycenter of seven short period (typically 2–3 hours) sub-dwarf (sdBs) eclipsing binary systems could indicate the presence of circumbinary objects: planets or brown dwarfs. Following our 246 new observations made between 2013 September and 2017 July using a worldwide network of telescopes, we found that some systems showed possible cyclical variation over the short term, but did not follow predictions. Only observations made over a very long timescale can resolve this and this is where amateur astronomers can make a significant scientific contribution. Full details of our paper entitled: "The quest for stable circumbinary companions to post-common envelope sdB eclipsing binaries? Does the observational evidence support their existence?" can be found in the March 2018 *Astronomy and Astrophysics Journal* freely available via the arXiv portal (https://arxiv.org/abs/1711.03749).

RZ Cassiopeiae: Light Curve and Orbital Period Variations

Geoff Chaplin
107 Clifton Street, London EC2A 4LG, United Kingdom; geoff@geoffgallery.net

Abstract   Recent electronic observations have shown that amateurs can obtain very high quality data from modest equipment. This talk shows several such observations and shows how they can be used to determine the type of eclipse, re-evaluate historical visual data, and calculate accuracy of times of minimum eclipse, and looks at the variation of the period and possible causes.

Williamina Paton Fleming's "Un-named" Variables and the AAVSO: A Scientific and Historical Perspective

Kristine Larsen
Central Connecticut State University, 1615 Stanley Street, New Britain, CT 06050; larsen@ccsu.edu

Abstract   Twenty years ago a *JAAVSO* article by Dorrit Hoffleit brought attention to the fact that fourteen of the nearly 300 variables discovered by Williamina Paton Fleming or her team at the Harvard College Observatory circa 1900 lacked permanent designations in the *General Catalogue of Variable Stars* (GCVS). Most of these stars have now received such designations. Since their discovery, much has changed in our understanding of these variables. An exploration of this evolution provides a valuable series of snapshots in time of the state of variable star astronomy over more than a century, and illustrates the ongoing and significant impact of the AAVSO and its observers on the field.

Cataclysmic Variables as Universal Accretion Laboratories

Christian Knigge
School of Physics and Astronomy, University of Southampton, Highfield, Southampton SO17 1BJ, United Kingdom; c.knigge@soton.ac.uk

Abstract   Cataclysmic variables (CVs) are numerous, bright and nearby, making them excellent laboratories for the study of accretion physics. Since their accretion flows are unaffected by relativistic effects or ultra-strong magnetic fields, they provide a crucial "control" group for efforts to understand more complex/compact systems, such as accreting neutron stars (NSs) and black holes (BHs). I will review recent work on CVs, which has revealed that these superficially simple systems actually exhibit the full range of accretion-related phenomenology seen in accreting NSs and BHs. Given this rich set of shared behavior, it is reasonable to hope that much of accretion physics is universal. CVs hold great promise in this context as observational testing grounds for attempts to model and understand this physics.

SN1987A and Connections to Red Novae

Thomas Morris
153 Ovaltine Court, Ovaltine Drive, Kings Langley, Hertfordshire WD4 8GU, United Kingdom; tom.s.morris@gmail.com

Abstract   I present a binary merger model for the progenitor of Supernova 1987A. A binary system initially consisting of 15 and 5 solar mass stars in a wide orbit, which merges some 20,000 years before core collapse, is able to explain many of the unexpected features of SN1987A. The common envelope phase gives rise to nova-like outbursts as primarily orbital energy is radiated from the common envelope. Such an outburst may explain the eruptions of V838 Mon, V1309 Sco, and perhaps the 1840s outburst of η Car. The 2001 to 2007 light curve of V1309 Sco observed by the OGLE project provides strong evidence for a merging binary within a common envelope.

ρ Cassiopeiae—an Update

Des Loughney
113 Kingsknowe Road North, Edinburgh EH14 2DQ, United Kingdom; desloughney@blueyonder.co.uk

Abstract   ρ Cas has been monitored by the author using DSLR photometry over the period 2007 to 2018 and the measurements are continuing. Following the outburst and fade in 2001–2002, which was thought to happen every 50 years or so, it was expected that the star would revert to its standard pattern of semiregular variations. The measurements between 2007 and 2013 seemed to confirm this. The star varied semi-regularly between 4.5 and 4.9 magnitudes. In 2013 an event occurred which heralded a new pattern of behavior. The star brightened to 4.3 and faded to 5 magnitude. A different pattern emerged which continued to 2018 when there was another mini outburst when the star brightened to 4.2. The brightenings, which are usually explained by mass ejections, are occurring more often and suggest that the dynamics of the star have changed.

American Medical Association Statement on Street Lighting

Mario Motta
19 Skipper Way, Gloucester, MA 01930; mmotta@massmed.org

Abstract The American Medical Association (AMA) has adopted an official policy statement about street lighting: use low blue LEDs. I am the principal author. I will show my presentation that I gave to the Illuminating Engineering Society (IES), who make the streetlight standards in the USA, and hope for change in their recommendations soon.

The LED street lighting that the industry had originally proposed and still suggesting is too harsh and bright for optimum safety and health. This report was adopted unanimously by the AMA House of Delegates at its annual meeting in 2016. It states that outdoor lighting at night, particularly street lighting, should have a color temperature (CT) of no greater than 3,000 K. Higher CT (4,000 K) generally means greater blue content, and the whiter the light appears.

A white LED at CT 4,000 K contains a high level (over 30%) of short wavelength, blue light. These overly blue harsh lights are damaging to the environment and have adverse human health effects. In some locations where they were installed, such as the city of Davis, California, residents demanded a complete replacement of these high CT street lights for lower CCT lighting. Cities that have followed the AMA recommendations and adopted 3,000 K or 2,700 K have seen much greater acceptance of LED lighting, and with much lower blue content which is better for human and environmental health, and reduces glare and is thus safer for driving.

The AMA has made three recommendations in its policy statement: First, the AMA supports a "proper conversion to community based Light Emitting Diode (LED) lighting, which reduces energy consumption and decreases the use of fossil fuels." Second, the AMA "encourage[s] minimizing and controlling blue-rich environmental lighting by using the lowest emission of blue light possible to reduce glare." Third, the AMA "encourage[s] the use of 3,000 K or lower lighting for outdoor installations such as roadways. All LED lighting should be properly shielded to minimize glare and detrimental human and environmental effects, and consideration should be given to utilize the ability of LED lighting to be dimmed for off-peak time periods."

Index to Volume 46

Author

Subject

PHOTOELECTRIC PHOTOMETRY [See PHOTOMETRY, PHOTOELECTRIC]

PHOTOMETRY

PHOTOMETRY, CCD

NOTES

Made in United States
Troutdale, OR
09/05/2023

12626192R10071